To my mother.
When I think of her, I always feel sorry and sad...

This book is an English and modified version of "새로운 양자물리와 초전도" written in Korean. (도서출판 홍릉, Hongreung Publishing Co. 30.01.2020, ISBN 9791156007289)

Copyright ©2020 Min Tae Kim

All right reserved
ISBN: 9798652813765

New Quantum Physics and Superconductivity

A physics novel for room-temperature superconductors

行憲 김민태[*1)]
Min Tae Kim, PhD

[*1)] The author of "Origin of Gravity and New Cosmos (ISBN 9781713042020, Kindle Direct Publishing 2019)".

Putting this book into the world...

One day, suddenly, an artificial intelligences (AI) has appeared and played Go to beat overwhelmingly Lee Se-dol. Though Lee won one game out of five, it was a shocking event in which some non-human intelligence surpassed humans. The thinking that humans will have to live as inferior beings and not the brightest creatures on Earth in the distant future, or not so far in the future, is reluctant. However, AIs are still works of human beings. Of course, they are not just works of human beings, but are the works of numerous scientists and engineers. Scientists and engineers here are not just those who are currently in business nor have worked for some time. AIs are the realizations of science and technology that has been studied and developed from the distant past to the present. If we think about it, AIs don't just exist in our real world. They exist only on so-called networks created by humans. They are like entertainers we can meet everyday on TV, but we need to turn on the TV. Of course, we can visit and meet them in person. Most of the time, however, the entertainers are only on TV or networks. Meanwhile, these intelligences are approaching us in human forms, robots. Yet the robot is as cute as a toddler, and we don't need to be alert. But if this robot becomes an adult... If it is stronger and more intelligent than us... We won't be able to control it. Uncertainty in human life will be very high. Horrible.

    The introduction was long. Now I want to tell you why I wrote this book. The AI is one of the implementations of the information of science and technology that has been accumulated from the distant past. Limited to science alone, The AI is a world of electric and magnetic signals, that usually wander in solid materials. It is a world of solid state physics.

It is no exaggeration to state that modern solid state physics has started with the duality of light, a wave and a particle. Quantum mechanics was born from the duality of light, and is in the core of modern physics. Einstein, famous for the theory of relativity, said, "God does not play dice." However, he won the Nobel Prize in Physics in 1921 for his theory of the photoelectric effect that contributed greatly to the birth of quantum mechanics. Since then, many quantum mechanical discoveries have been made, and solid-state physics has developed remarkably. There is no phenomenon that cannot be explained by quantum mechanics, from the flow of electricity, magnetic phenomena, etc. down to the behavior of subatomic particles. It looks like that. Of course, classical mechanics explained macroscopic phenomena well, but the micro-world could not be accurately explained without quantum mechanics. There are things, however, that even quantum mechanics has not yet solved, as there is no perfect theory in the world. One of them is superconductivity. The main reason for writing this book is to clarify the mechanism of superconductivity that quantum mechanics cannot solve.

Superconductivity in solids is a physical phenomenon in which electrical resistance of a conductor goes to zero at very low temperatures that we do not experience in everyday life. It also includes the magnetic phenomenon that the conductor completely repels magnetic fields. In 1957, John Bardeen, Leon Cooper, and John R. Schrieffer in the United States interpreted the superconducting phenomena in terms of quantum mechanics with their BCS theory, which borrowed the first letters of their family names. For this, they were jointly awarded the Nobel Prize in Physics in 1972. This theory explained the superconductivity of metals at very low temperatures (below 30 K, -243°C), but it did not go well along with oxide based high temperature superconductors. In the BCS theory, paired electrons called Cooper pairs are the main charge carriers in superconductors. However, it is impossible to explain high temperature superconductivity with Cooper pairs. There appears to be no need to insist on the BCS theory that is different from the theory of normal electrical conductivity and that does not fit with high temperature superconductivity. However, no solid theory has yet appeared to replace this theory and to cover high temperature superconductivity and electrical conductivity at room temperature simultaneously.

Both superconductivity and electrical conductivity in solids are the same flow of electricity. It is a "flow" of electrical energy. If an electric current is a flow of energy, charged particles called electrons does not need to deliver the energy directly. No one has ever seen these charged particles flowing through conductors. There seems to be no way to observe them. There are no particles that carry the elastic energy of solids. The relative positions of the constituent atoms or molecules propagate in the form of waves. Electricity also does not involve particles that carry its energy. In this regard, we accessed electricity and searched for the nature of superconductivity. To do this, a new atomic model was set up and electromagnetism was newly defined. We wanted to create a new paradigm to explain electromagnetic phenomena in solids. We did thought experiments for new solid-state physics and recorded the experimental results in this book. It won't be easily understood for the readers out there. It appears to an absurd but is a fun challenge for new physics. I appreciate all the people who helped me publishing this book.

Introduction

In the modern civilized society, one cannot live without electricity. We saw what the world without electricity would be like during the California blackout of early 2001. According to an estimation, if electricity is shut down immediately and forever, two thirds of the US population would be lost. The question, what is electricity in solids, is the main topic of this book. How is electricity generated and how does it flow? Dealing with this is the main job of solid-state physics. We imagined the flow of electricity without electrons. Are there any electrons flowing through materials? Has anyone ever seen electrons flowing through copper wires? In fact, there is no way to see the "free" electrons flowing through conductors, including copper wires. We only indirectly predict what would happen under the assumption that electrons flow freely in conductors, and think that there are free electrons if the prediction appears to be a good fit. One of the representative things of this prediction is the magnetic field accompanying the electric field. The flow of free electrons creates an electric field and induces a magnetic field. But let's think of electromagnetic waves represented by light. An electric field is created and a magnetic field is formed without electrons in light. It is that electricity can flow without free electrons.

The absence of free electrons means that electrons in all materials including conductors belong to something else. Electrons must of course be with protons. In modern physics, the number of electrons and protons in an atom is always the same, but they are not involved in the construction of the atom in pairs. All the protons are centered in the core called the nucleus, while the electrons are roaming around the nucleus. Why can't the electrons and protons come together? Wouldn't it be harmonious if they form atoms in pairs? With this idea, we made a new atomic model. In this model, the electrons and protons are always in pairs, and the neutrons bind them together to form various elements. The triplet of electron-proton-neutron (like deuterium) was considered to be the elementary component in the atom except for the hydrogen atom. We imagine that the proton is a grape seed, the electron is pulp and the neutron is like a raisin in our atomic model. In addition, the electron and proton in one pair are not always present as separated, but they combine to become a neutron and the neutron is converted back to the electron-proton pair in dynamic equilibrium.

If this triplet is the building block of the atom, how will the atom build up this building block? If there is no biased force from outside, it will accumulate in a spherical form. A smaller sphere will grow to a bigger one as the triplets pile up on the surface of smaller one. Once a layer of the triplets has been completed to form a hollow sphere, the triplets will pile up sequentially to build an additional layer on top of it. Heavy atoms have multiple such layers, the number of such layers means the period of the periodic table of elements. At the end of each period, atoms have a perfect spherical form. These are the atoms of Group VIIIA, but at the beginning of the next period, the atomic surface is uneven and thus unstable. Unstable atoms will accommodate other unstable atoms to be stable. They make chemical bonds.

Lowering temperatures reduce the interatomic distance and the distance between the outermost triplets of neighboring atoms in materials. At the critical distance, the attraction is sufficient to form an interatomic chemical bond. A chemical bond can be a covalent bond, an ionic bonds, or a metallic bond. A metal conductor is mainly composed of metallic bonds, but in a conductor such as an oxide based superconductor atoms are covalently bound. While some triplets form the skeleton of a solid by primary chemical bonds, some triplets, which are not participated in the primary bonds, form secondary bonds with other triplets of the adjacent atoms. We may think hydrogen bonds or van der Waals forces as a type of secondary bond. We may also think of the alignment of atomic magnetic moments in solids to be a kind of secondary bond. Secondary bonds are produced and destroyed constantly in thermally vibrating solids. The density and bonding force of secondary bonds vary with external temperature and pressure, and the electromagnetic properties of conductors change correspondingly. Some secondary bonds can determine the electromagnetic properties of solids. We call them specifically secondary electromagnetic bonds throughout this book. At room temperature, the bond density is so low that the constituent atoms do not form an electromagnetically regular array. When temperature is lowered to a critical temperature, the density of the secondary bonds is reached to a critical value where an electromagnetic phase transformation (solidification) occurs in the interiors of a conductor, The electromagnetic deformation of the secondary bonds in the conductor is not allowed below the critical temperature.

Electricity does not flow inside the conductor, making it a perfect insulator. For some conductors, however, electricity can flow only through the surface layers via the interaction with the solid vacuum without resistance. This surface current is the current of superconductivity.

　This book reconsidered the meaning of interatomic primary and secondary bonds, based on the newly conceived atomic model. On this base, superconductivity, especially high temperature superconductivity, was addressed in terms of the reaction of electromagnetic secondary bonds to external conditions and the resulting electromagnetic properties of solids. This book is dedicated to room temperature superconductors, the dream material, that will be discovered someday.

Contents

I. Solid vacuum and energy 13
  1.1. Structure of the solid vacuum 14
  1.2. Source of energy 17
  1.3. Birth of electron and positron 19

II. Birth of matter and new atomic model 29
  2.1. Proton and neutron 30
  2.2. History of atomic model 39
  2.3. Hydrogen and helium 45
  2.4. New hydrogen atom model 52
  2.5. Atomic structure in new atomic model 57

III. Chemical bond 69
  3.1. Energy level and stress field around crystal defects 71
  3.2. Interatomic bonding 79
  3.3. Hydrogen bond 90
  3.4. Van der Waals force 104

IV. Electromagnetic property of solids 115
  4.1. Electromagnetism 116
  4.2. Dielectric property of solids 123

4.3. Magnetism in solids   131
4.4. Theory of magnetism   152

V. New paradigm of electromagnetism   159
5.1. Electric current in solids   160
5.2. Non-electron model of conductivity   165
5.3. Magnetism in the new paradigm   182
5.3. Maxwell equations in the new paradigm   187

VI. New paradigm of superconductivity   207
6.1. Superconductivity and the BCS theory   209
6.2. Superconductivity in the new paradigm   214
6.3. High temperature superconductors   230
6.4. Unified theory of superconductivity   248

VII. For room temperature superconductors   263

References   277

I. Solid vacuum and energy

In the first edition of "Origin of Gravity and New Cosmos, 2019 (ISBN 9781713042020)", the author sought to find the origin of the mysterious force of gravity in the vacuum (in the background of the universe). The vacuum is not "nothing" but consists of a medium with a much higher density than any matter we know of, and on the basic premise that it has zero-point energy, physical phenomena, such as the propagation of light, gravity and other cosmological phenomena were newly interpreted. Matter is not separated from the solid vacuum, but is energy stored in the vacuum, causing it to deform as much as the energy stored. A mass is a deformation of the solid vacuum, which is also a lump of energy.

In this chapter, we set up a hypothesis about the structure of the solid vacuum and how energy is stored in there, if the vacuum is a solid with no energy or minimum energy. Thereon we try to provide basic ideas for establishing a new theory of superconductivity that is the main topic of this book.

1.1. Structure of the solid vacuum

In "Origin of Gravity and New Cosmos"(ISBN 9781713042020), the vacuum was defined to be a very dense medium, which does not convert into energy, and matter is nothing but the (vibration) energy stored in it. When energy is input to the solid vacuum, electrons and protons (or neutrons) are created, which make various atoms from the combination of the electrons and protons/neutrons, and molecules composed of various atoms.*2)

What could be the crystallographic structure of the solid vacuum free of energy (or with only a slight energy called zero-point energy)? A solid can have a face centered cubic (fcc) or hexagonal close packed (hcp) crystal structure at low temperatures, as shown in Figure 1, when the atoms are not biased. This is because these structures are the most stable energetically. Here we consider the minimum unit that makes up the solid vacuum. This may be a point particle or lattice point. This unit is different from the neutron or neutrino. Assuming that the solid vacuum is an energy-free medium made of such lattice points or particles, it will presumably form a regular lattice structure as shown in Figure 1. How can we infer that the solid vacuum forms an ordered lattice?

---

*2) The discussion can start with the elementary particles called quarks. Particle physics states that the combination of quarks gives baryons, and protons and neutrons are baryons with three quarks combined. However, quarks that make up protons or neutrons are not observed directly, but indirectly identified from the force mediating gluons. Thus, quarks are not existing particles, but interactions between the elementary lattice points of the solid vacuum mediated by gluons, kinds of springs.

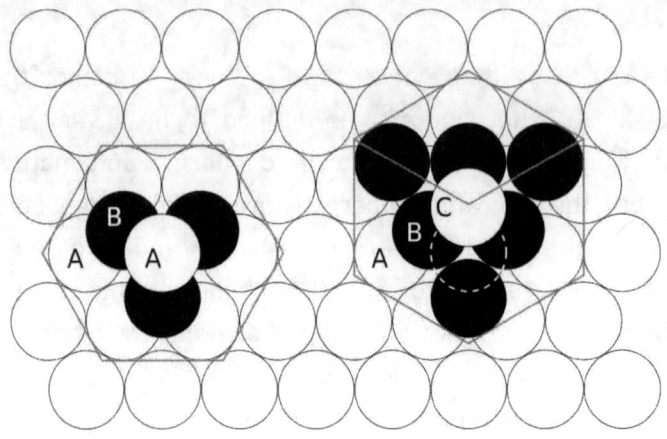

Figure 1. Stack of spherical particles that form a hcp lattice (left) and fcc lattice. They are differentiated according to the arrangement stacked from the white base layer (image from wikipedia.org).

Light reveals the structure of the solid vacuum?

Light travels through the solid vacuum. The pattern of the propagation of light gives some ideas about the vacuum structure. For example, it is seen from Doppler's effect or the behavior of light around a massive star that the solid vacuum is compressed or elongated. The change in the wavelength of light along the compressed solid vacuum or the deflection of light in a gravitational field suggests the regularity of the solid vacuum. We can't see the Sun with our naked eyes, but if we give it a moment, we see that the Sun does not radiate evenly in all directions. The same is true for the image of stars. Stars have a common image imprinted with the star shape (☆) since ancient times. The image of stars was not recognized with telescopes, but with

Figure 2. Bright stars observed: (top left) Canopus at 310 light years; (top right) Sirius at 8.6 light-years; (bottom left) Lyre Alpha, Vega at 25 light-years; (bottom right) Arcturus at 37 light-years (Alpha star in Zodiac) (image from wikipedia.org).

naked eyes. As seen in Figure 2, some bright stars actually observed appear to propagate quickly or intensively in particular directions. There are directions for light to propagate preferentially. There seems no other ways, but we should relate this behavior of light to the regular lattice structure of the solid vacuum. Based on the assumption that the solid vacuum has a energetically stable lattice structure, we would like to propose a new model for atoms that make up ordinary matter.

1.2. Source of energy

We know well matter is energy from Einstein mass-energy equivalence ($E = mc^2$, E is energy, m is mass, c is the speed of light). If the solid vacuum can not be converted to energy, where comes the energy to be matter in the solid vacuum? There are things only existing in the form of energy. They are electromagnetic waves, represented by light. The mass of light is zero, but has energy. Light reacts with matter to be absorbed or emitted. Light is an electromagnetic wave, namely it is a wave. Its energy is given as the product of its frequency and the Planck[*3)] constant. In the book, "Origin of Gravity and New Cosmos", we showed that the propagation of light in the solid vacuum can be understood in the same way as shear elastic waves traveling in solids. The velocity of a shear elastic wave in a solid is given as

$$v_s = \sqrt{\frac{G}{\rho}} \quad \text{---} \quad (1.1).$$

[*3)] Max K.E. Ludwig Planck (4.1858 - 10.1947) was a German theoretical physicist. He won the Nobel Prize in Physics in 1918 for the discovery of energy quanta. As one of the founders of quantum theory, he revolutionized the understanding of the atomic and subatomic physical world. To honor him, Germany changed the name of the Royal Wilhelm Institute, which served twice as director, to the Max Planck Institute in 1948.

Here, G is the shear modulus and ρ is the mass density. On the other hand, the speed of light c from mass-energy equivalence is

$$c = \sqrt{\frac{E}{m}} = \sqrt{\frac{G_V}{\rho_V}} \quad \text{---} \quad (1.2),$$

where V is the volume. $\rho_V = m/V$ is the density of the solid vacuum. $G_V = E/V$ is the shear modulus of the solid vacuum. The vacuum density is unknown, but it is calculated as 4.64×10^{100} GPa by substituting the theoretical Planck density of 5.16×10^{94} g/cm³ for ρ_V. It is unimaginably larger than the shear modulus of 440 GPa or 470 GPa of diamond, which is known to be the hardest in the world.[1]

Light and neutrinos - the source of matter and energy

Where comes light propagating through the solid vacuum at the speed of light? In our daily lives we can see that light comes from electric lights or fires. But most of the light energy comes from the Sun. The Sun's light energy is produced by nuclear fusion, the fundamental process of combining protons and electrons of hydrogen to yield neutrons. Other stars in the universe go through the same process and glow. The fusion reaction occurring in the interiors of the Sun is known as the p-p chain reaction.[2] However, the energy produced by this process is emitted not only in the form of light but also of neutrinos. Neutrons produced by the solar fusion reaction reach Earth every second 65 billion particles per cubic centimeter. The fusion reaction is basically the fusion of the positively charged protons and negatively charged electrons of hydrogen atoms. If we assume that a proton is the composite of a neutron and a positron, in this fusion reaction, the positron combines with the electron (and the proton becomes a neutron), to emit electromagnetic waves and neutrinos. From this process, we may also imagine a reverse reaction in which light and neutrinos convert to electrons and protons, and atoms, molecules, etc.

1.3. Birth of electron and positron

According to the standard model of the universe, the universe expands rapidly after 10^{-37} seconds of the Big Bang, eliminating almost all the irregularities.[3] The remaining irregularities caused further expansion by quantum fluctuations (or vacuum fluctuations).[4] A quantum fluctuation is the temporary variation of energy in a point in space, according to the Heisenberg[*4] uncertainty principle. In quantum theory, short-lived "virtual" particle-antiparticle pairs in the vacuum between two interacting particles are created with energy corresponding

---

*4) Werner Heisenberg (12.1901 - 2.1976) was a German physicist. One of the pioneers of quantum mechanics. In 1927 he published a historical paper on the principle of uncertainty and won the Nobel Prize in Physics for his contributions to the birth of quantum mechanics in 1932. He also made achievements in the field of hydrodynamics, atomic nuclei, ferromagnetism, and theory of cosmic rays. He was a member of Germany's first nuclear reactor and was a senior scientist at the Nazi nuclear weapon project during World War II.

to time in the time-energy relationship based on the Heisenberg uncertainty principle,[*5)] and then vanish after that time. A hypothetical electron-positron pair acts as an electrode to rearrange the electric field (the electromagnetic field around the electron). This rearrangement makes the electromagnetic field applied to the outside weaker than it would have been in the vacuum. This is vacuum polarization.[5] Light propagates in the solid vacuum as a positive and a negative electric field alternately appears and disappears. In this process, if the energy of light is absorbed via vacuum polarization, the virtual electron-positron pair will really come into being. This means that electrons and positrons can be created from light.

Electron and positron originated from light

When an electron merges with a positron, electromagnetic waves called gamma rays are generated. Could electrons and positrons be the special forms of light? If an electromagnetic wave is twisted to form a circular standing wave, as shown in Figure 3, the shape of the electric field inside and outside the circle will be distorted, and due to this difference the circular standing wave will appear to be charged. If the negative part of the electric field faces outward, it becomes an electron, and the positive part

---

[*5)] When measuring two physical quantities related to each other simultaneously, there is a physical limit to the accuracy between the two. According to this principle, the position and momentum of a particle cannot be measured simultaneously, and the higher the accuracy of position, the lower the accuracy of momentum.

facing outward will make a positron. The spin of electron or positron can be understood as the rotational movement of the circular electromagnetic wave.

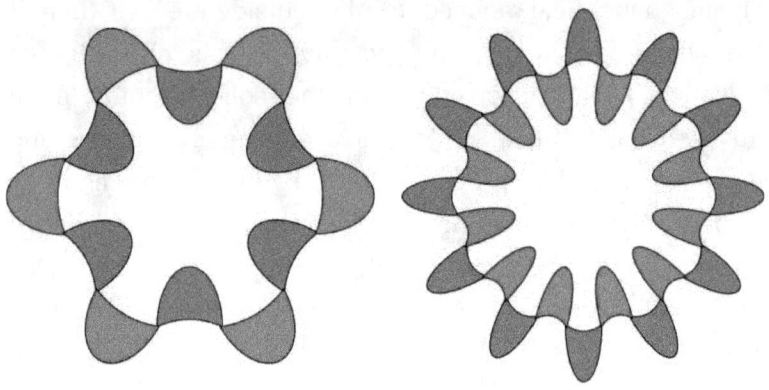

Figure 3. Model of electron or positron. When an electromagnetic wave called light form a circular standing wave, it becomes an electron or a positron. The electron energy is higher if the frequency of the standing wave is higher. Since the frequency of a standing wave is quantized, the electron energy is also quantized.

If electrons were generated by quantum fluctuations after the Big Bang, where did the same number of positrons go? We know of positive beta decay, a type of radioactive decay, in which a proton becomes a neutron in the atomic nucleus of a radioactive element, releasing a positron and electron neutrino.[6] A typical example is the process in which ^{23}Mg decays to ^{23}Na. In respect of positive beta decay, we may think that a proton is made of a neutron and a positron, the neutrons contained in an atom can be regarded as transients formed by losing the proton energy. A neutron star, one of the supernova remnants, is believe to lose its energy to become a black hole, which eventually dies out. So even the neutrons produced by fusion may die out to leave only the (virtual) lattice of the solid vacuum free of energy. If a positron is captured at the vacuum lattice point and become a proton, it will not be easy to find free positrons in the universe.

We think of a scenario where light forms electron-positron pairs through the phenomenon called vacuum polarization, in which positrons are trapped in the solid vacuum to be protons. If a positron is trapped in the vacuum lattice, this lattice point should be negatively charged so that a positron will be captured to be energetically stable. According to a paper published in 2007,[7] the outside of a neutron is negatively charged, the outer center has a positive charge, and the core has a negative charge. This may mean that the vacuum lattice points are negative in charge and positrons easily combine with these negative lattice points to form protons and then fuse with electrons to become neutrons. Let's see how electrons and positrons are found and what they look like in modern physics.

Discovery and properties of electrons

An electron (denoted by e⁻) is known to be a particle of zero diameter and volume. These particles cannot be observed optically and indirectly predicted and observed. In the 1870s, Crookes[*6] confirmed that beams generated in a high vacuum cathode tube were attracted to the cathode

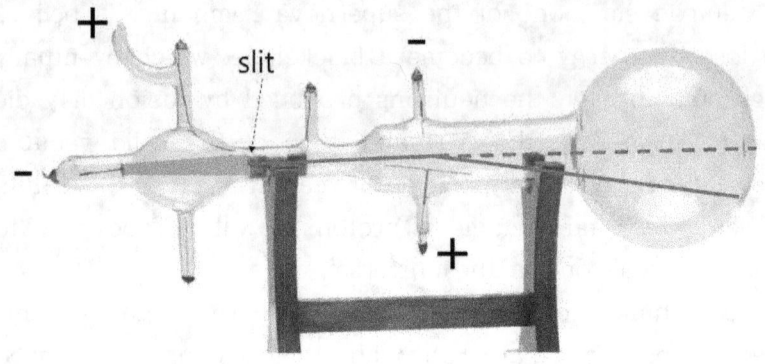

Figure 4. A beam from the cathode at one end of the glass vessel passes through the slit and then bends between the two middle electrode plates (J.J. Thomson).

and deflected in a magnetic field. The beams were called as negatively charged matter.[8] In 1890 Schuster[*7)] installed two metal plates parallel to the path of the beams and applied a voltage across them, and calculated the charge-to-mass ratio from the beam deflection. However, this value was more than 1,000 times higher than expected, so it did not attract attentions.[9] In 1892 Lorentz[*8)] proposed that the

---

*6) William Crookes (6.1832 – 4.1919) was a British chemical physicist. Discovered thallium in 1861.
*7) Arthur Schuster (9.1851 – 10.1934) was a British physicist from Germany.
*8) Hendrik A. Lorentz (7.1853 – 2.1928) was a Dutch physicist. He discovered the Zeeman effect, the splitting of the emission spectrum of an atom or molecule into multiple lines by an external magnetic field. He also derived a transformation equation that is important to the theory of special relativity. He was awarded the Nobel Prize in Physics in 1902 for the discovery of the Zeeman effect and his theory on the effect together with Pieter Zeeman (5.1865 –

mass of this particle was due to electric charges,10 and in 1896 J.J. Thomson[*9)] et al. found that the cathode beams are composed of particles whose charge-to-mass ratio is irrelevant to the cathode material.11 In 1900, beta rays from radium were bent in an electric field, and the charge-to-mass ratio was found to be the same as that of the beams from the cathode tube, so that the electron became a member of atomic constituents.12 Although electrons were recognized as particles, in 1924 de Broglie[*10)] published a hypothesis that matter, including electrons, can be interpreted as waves, like light. 1927 G.P. Thomson[*11)] confirmed the de Broglie's hypothesis by discovering that interference occurs when an electron beam passes through a thin metal film.

The electrons studied and observed so far, especially in quantum mechanics, are the first generation of lepton and elementary particles. The electron mass is 9.109×10^{-31} kg or $0.511 MeV/c^2$, 1/1836 of the proton mass. The amount of charge is -1.602×10^{-19} coulombs and the volume is zero.13 Quantum mechanically, the electron has its own spin of ½, a fermion[*12)]. Any two electrons cannot have the same

---

10.1943).
*9) Joseph J. Thomson (12.1856 – 8.1940) was a British physicist. He discovered electrons and identified one of the particles that make up an atom for the first time. In 1906 he won the Nobel Prize in Physics for his work on electrical conductivity.
*10) Louis V.P. Raymond de Broglie (5.1892 – 3.1987) was a French physicist. He introduced the hypothesis of matter wave in his PhD thesis in 1924, and won the Novel Prise in Physic 1929 after his theory was experimentally proved in 1927.
*11) George P. Thomson (5.1892 – 9.1975) was a British physicist. He was awarded the 1937 Nobel Prize in Physics for discovering the interference of electrons in crystals.

quantum state. Electrons emit or absorb photons when they are accelerated. Electrons can be produced by negative beta decay of radioactive isotopes (the neutron releases an electron and a neutrino and become a proton) or via high-energy collisions, such as when cosmic rays enter the atmosphere. When electrons and positrons collide, they disappears by generating gamma rays.*13) According to the uncertainty principle of energy, the diameter is up to 10^{-18} m and the classical diameter is 2.8179×10^{-15} m. The lifetime is estimated to be 6.6×10^{28} years.14

Discovery and properties of positrons

In 1928 Dirac*14) presented an electron model comprising relativity and symmetry in quantum mechanics of electromagnetism. It is the Dirac equation, from which quantum states with abnormal negative energies were predicted.15 In order to solve this problem, Dirac proposed a vacuum model consisting of an infinite sea of particles

---

*12) It is named after Enrico Fermi (9.1901 - 11.1954), a Italian American physicist. He developed the world's first nuclear reactor, the Chicago Pile 1 and won the Nobel Prize in Physics for his work on atomic nuclei in 1938.

*13) Highly energetic electromagnetic waves resulting from rapid fission processes such as the case with nuclear bombs. The amount of energy is about 10 keV.

*14) Paul Dirac (8.1902 - 10.1984) was a British theoretical physicist. In 1933 he won the Nobel Prize in Physics along with Schrödinger. He is respected as one of the greatest physicists of the 20th century. He made major contributions to the development of early quantum mechanics and quantum electrodynamics.
The Dirac equation describes the behavior of fermions and predicts the presence of antimatter.

with negative energies (the Dirac sea) in 1930.[16] In this model, the electron could not have negative energy and the presence of a new particle, a positron, was predicted. Positrons had been observed by several researchers, but the existence was not clearly imprinted and they became finally positrons in 1932 by Anderson*[15] who confirmed the existence.[17]

The positron has a charge of +e and a spin of ½ as with the electron. When collided with an electron, it disappears upon releasing gamma rays. Protons are produced via radioactive decay, such as beta decay of radioactive isotopes, or when highly energetic photons react with atoms in materials to produce electron-positron pairs.[18] Accelerated electrons in the cloud's strong electric field create gamma ray flashes, and positrons have been found in the flashes.[19] Potassium isotopes ^{40}K have a long half-life and are also present in our body, so an adult of 70 kg produces 4,000 positrons per second through beta decay, and they soon decay to gamma rays by colliding with electrons.[20]

The electron and positron are a perfectly symmetrical particle-antiparticle pair that combine to form light. Reversely, we can hypothesize that electrons and positrons are generated from light. The behavior of electrons in the solid vacuum can be confirmed by the cathode tube experiment as shown in Figure 4, but the nature of free electrons in a conductor is unknown. When light is irradiated on the surface of a conductor, it is not known

---

*15) Carl D. Anderson (9.1905 - 1.1991) was an American physicist. He was awarded the Nobel Prize in Physics in 1936 for his discovery of positron.

whether electrons emitted via the photoelectric effect*16) actually come out of the conductor or whether electrons and positrons are generated from the incident light, so that the positrons remain in the conductor and only the electrons are emitted. The concept that an electric current is the flow of free electrons is ambiguous.

---

*16) The photoelectric effect is a phenomenon in which electrons are emitted when light below a certain wavelength (above a certain frequency) is shed onto the surface of metals. Einstein won the Nobel Prize in Physics for his theory on the photoelectric effect in 1921.

II. Birth of matter and new atomic model

If mass-energy equivalence holds, the source of energy would be the things with very little or no mass, such as light or neutrinos. As electrons and positrons combine to become light, light can be divided into electrons and positrons. When positrons are captured by the solid vacuum lattice, they turn into protons. Protons combine with electrons to form hydrogen atoms, which then undergo fusion to form neutrons, which facilitate forming heavy elements. During this process, the positron released from the proton combines with an electron again to emit electromagnetic energy and to disappear. The smallest mass of hydrogen has the highest energy density, and the release of energy results in the birth of new heavier atoms. Atoms consist of protons, electrons, and neutrons. The neutrons in stable atoms are energetically stable than the protons. This chapter summarizes what current physics tells us about the protons and neutrons that make up atoms, and presents a new theory regarding to the inner structure of the atoms heavier than the hydrogen atom, and discuss its validity.

2.1. Proton and neutron

Discovery and property of protons

In 1886 Goldstein[17] discovered that an anode beam from the gas discharge tube, shown in Figure 5, was of positively charged particles, but did not identify it as a group of single particles such as protons. After Rutherford[18] discovered nuclei in 1911, he found in 1917 that the hydrogen nucleus is also present in other nuclei. This is the discovery of protons. The proton is a stable particle that is a member of the nucleus. But, unlike free neutrons, free protons do not convert to other particles. Protons with high energy and velocity make up 90% of cosmic rays. The SK station[19] in Japan found that the minimum proton lifetime is 6.6×10^{33} years, being decomposed into anti-muons[20] and neutral pions, and after 8.2×10^{33} years,

---

[17] Eugen Goldstein (9.1850 - 12.1930) was a German physicist.
[18] Ernest Rutherford (8.1871 - 10.1937) was a British physicist. He introduced the concept of half-life of radioactive elements and discovered the radioactive element radon, and identified the difference between alpha and beta rays. In 1908, he won the Nobel Prize in Chemistry. He presented an atomic model called Rutherford's model in 1911, and separated protons in the nuclear reaction between nitrogen and alpha particles in 1917.
[19] Super-Kamiokande. Neutrino Observation Research Center located in Mt. Hida's Mozumi mine in Japan. It is 1,000m underground. The station was established to observe neutrinos during proton decay and from the Sun.
[20] A muon is an elementary particle with the same charge -e and spin ½ as an electron, but the mass is very large (~207 times that of electron). Muons are unstable particles (the lifetime 2.2 μs) that only slowly decay by weak interactions. The decay produces electrons and two neutrinos. An anti-muon is

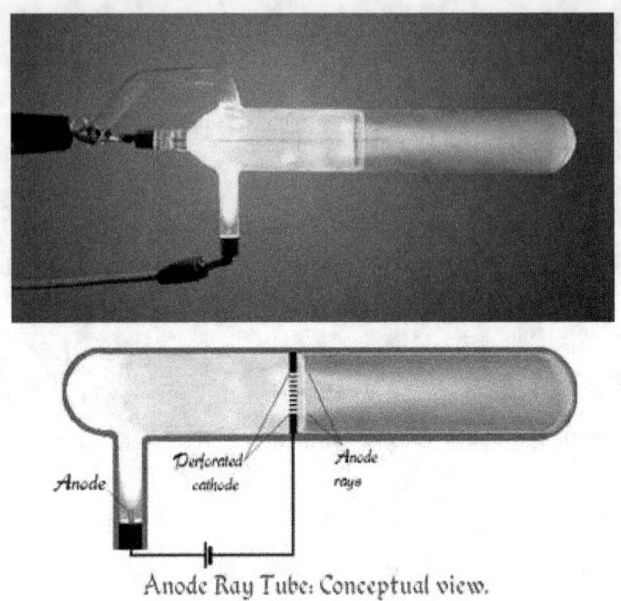

Figure 5. A beam from the anode passes through the holed cathode and becomes a pink beam (image from wikipedia.org).

they decompose into positrons and neutral pions.[*21)21]

A proton consists of three quarks in the modern standard model, as shown in Figure 6. The quark rest mass is about 1% of the proton's mass. The rest mass comes from the binding energy of quantum chromodynamics[*22)],

---

ab anti-particle with a positive charge.

[*21)] There are three types of pions: π^0, π^+, and π^-. They consist of quarks and anti-quarks, a kind of meson. Charged pions, π^+, and π^- have a lifetime of 2.6×10^{-8} seconds and neutral pions π^0 of 8.4×10^{-17} seconds. Charged pions decay into muons and muon neutrinos, and neutral pions convert to gamma rays. Pions are created via energetic particle collisions or when cosmic rays collide with Earth's atmosphere.

[*22)] A gauge theory explaining strong interaction. This theory

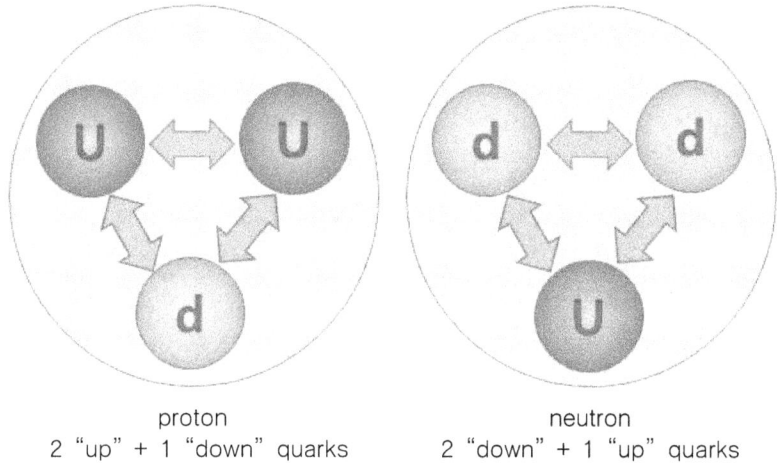

proton
2 "up" + 1 "down" quarks

neutron
2 "down" + 1 "up" quarks

Figure 6. Proton (left) and neutron (right) models consisting of three quarks.

being the quark kinetic energy and the gluon energy that combines the quarks. The proton mass is 80-100 times larger than the total mass of the quarks. The gluons have no mass but have energy of quantum chromodynamics.

The proton mass is 938 MeV/c^2, to which the quarks contribute only 9.4 MeV/c^2.[22] The charge radius of a proton is 0.8768 fm (0.8768×10^{-15} m) obtained from electron scattering tests and studies on the atomic energy levels of hydrogen and deuterium.[23] Recent measurements, however, differ slightly from this value, and the proton diameter remains a mystery and quantum electrodynamics is being reviewed.

---

explains the types and properties of quarks and gluons.

Neutrons in various forms

In 1928 Bothe*23) observed that when beryllium (atomic number 4) atoms collide with alpha particles, they release highly penetrating and electrically neutral particles. At first these were thought to be highly energetic gamma rays.24 But Chadwick*24) found that the ionization effect was too strong to be seen as electromagnetic waves such as gamma rays. In 1932 he measured the energy of the charged particles that bounced back by exposing this unknown beam to a group of elements such as hydrogen or nitrogen, and concluded that the beam is made of neutral particles whose mass is similar to that of proton.25 Neutrons were finally discovered.

The neutron is a hadron composed of quarks in the standard model and is a kind of baryon composed of three quarks (2 down quarks and an up quark) as shown in Figure 6. The total mass of the quarks is 12 MeV/c^2. As with the proton, the quarks are tied up with gluons by strong interactions. A free neutron has the mass of 1.675×10^{-27} kg (939.6MeV/c^2),26 the average diameter of 0.8×10^{-15} m (0.8 fm),27 and is a fermion with a spin of ½.28 Since the neutron has no charge, the mass cannot be

---

*23) Walther Bothe (1.1891 - 2.1957) was a German physicist. He worked on the wave-particle duality of radioactive rays and cosmic rays, the Compton effect, and nuclear reactions. For this, he won the Nobel Prize in Physics in 1954 with M. Born (12.1882 - 1.1970, German British physicist).

*24) James Chadwick (10.1891 - 7.1974) was a British physicist. In 1932 he won the Nobel Prize in Physics for the discovery of neutrons. Supported by the United States, he participated as a key figure in the development of nuclear weapons.

measured by mass spectrometry. Indirectly, the mass of a deuterium nucleus (with one proton and one neutron) is subtracted from the mass of proton. This includes the binding energy, which is the sum of the gamma ray energy (0.7822 MeV) released when a neutron is captured by a proton (zero neutron energy) and the rebound energy of the deuterium nucleus (about 0.06% of the total energy). The gamma ray energy can be measured by x-ray diffraction techniques and the most accurate neutron mass measured is 1.008644904 (14) u*25) in 1986.29

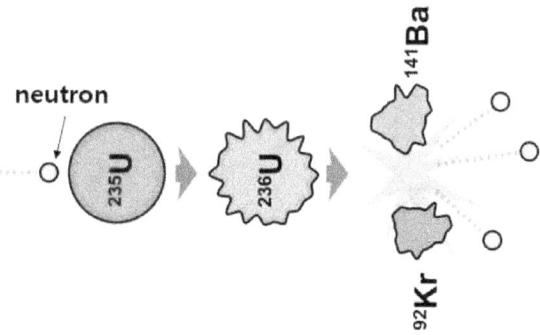

Figure 7. Nuclear fission induced by free neutrons (image from wikipedia.org).

The proton and neutron show almost the same behavior in the nucleus. In nuclear fission, the atoms that become unstable by absorbing neutrons (e.g. ^{235}U) decompose into smaller ones and emit other neutrons, as shown in Figure 7. Free neutrons are unstable and decays into protons,

---

*25) unified atomic mass unit, $1.660539040×10^{-27}$ kg

electrons, and anti-neutrinos within 15 minutes (881.5±1.5 seconds). This process is called beta decay which behaves according to the weak interaction. When a free neutron decay, the energy is 0.782343 MeV, in which the electron energy is 0.782 MeV and the rest is the neutrino energy.[30]

The neutron has no electric charge, but has a magnetic moment because there are quarks in the substructure and the internal charge distribution is not uniform.[31] The magnetic moment was first measured in 1940 using a kind of magnetic resonance method.[32] The magnetic moment of neutron can be regarded as the vector sum of the magnetic moment of the three quarks plus the sum of the orbital magnetic moments arising from the behavior of the quarks in the neutron.[33]

According to a paper published in 2007,[34] the outside of a neutron is negatively charged, the central part is positive, and the core is again negatively charged.*[26] This charge distribution explains that the location of magnetic polarization of the neutron is in the opposite direction to the spin angular momentum vector. Unlike the proton, the magnetic moment of the neutron is similar to that of the particle with a negative charge and can be said to have a magnetic field because the radius of negative charge distribution is broad. According to the standard model of particle physics, the negative and positive charges are not uniformly distributed in the neutron and there is a

---

*26) This means that the vacuum lattice point is negatively charged and a positron is trapped at this negative lattice point to form a proton. When fused with an electron, it becomes a neutron. From the classical point of view, if the outside is negative, it has affinity with the proton. However, the attraction of protons and neutrons does not involve electric charges.

constant electric polarization moment, but this value is too small to measure.35

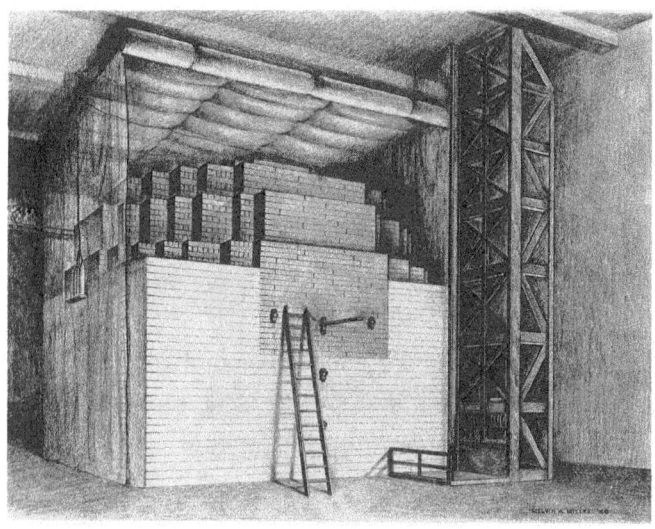

Figure 8. The world's first nuclear reactor, the Chicago Pile-1 (CP-1) (M.A. Miller)

Composition of the atomic nucleus

Protons together with neutrons form atomic nuclei. The protons in the nucleus are mutually strongly repulsive, so the neutrons suppress this repulsion to stabilize the nucleus.*27) The specific structure of nucleus is not yet

---

*27) In the regime of the new vacuum paradigm the neutrons in stable atoms are energetically stable, resulting from nuclear fusion induced by very high pressures inside a star where the electron in a hydrogen atom is fused with the proton. The potential energy of the electron toward the proton is released

clear. The number of protons in an atom is equal to the atomic number Z. The neutron number is N and thus the atomic mass is $Z+N = A$.[28] In the nucleus, the protons and neutrons are bound by the nuclear force, and the neutrons are necessary for the stabilization of nucleus. Neutrons are produced by nuclear fission or fusion. After the discovery of the fission reaction in 1938, it was found that a fission reaction produces free neutrons that caused other fission reactions. The first to use this chain reaction is the nuclear reactor CP-1[29] in Figure 8 and the nuclear weapon Trinity.[30]

In 1920, Rutherford suggested that the nucleus was composed of protons and electrically neutral particles, and the electrons were bound to the protons in some way because the protons and electrons are released simultaneously during beta decay.[36] In the 1920s, physicists believed that the atomic nucleus consisted of protons and "nuclear electrons." This is quantum-mechanically contradicted by Heisenberg's uncertainty principle, and in 1929 it was demonstrated that electrons could not be held

---

as electromagnetic waves and neutrinos, yielding a neutron.

[28] $B = N-Z$ is the neutron excess. The state of excess neutrons is assumed to be energetically similar to that of the cold solid vacuum in the new vacuum paradigm.

[29] Chicago Pile-1, The world's first artificial nuclear reactor, developed in 1942. It was built at a metallurgical engineering laboratory in the University of Chicago's as part of the Manhattan Project for producing atomic bombs during World War II. On December 2, 1942, the first nuclear chain reaction began under the supervision of Fermi.

[30] Trinity was the codename of the first nuclear explosion. As part of the Manhattan Project, US forces exploded the nuclear bomb at Jornada del Muerto, 56 km southeast of Socorro, New Mexico, at 5:29 on June 16, 1945.

quantum-mechanically in the nucleus.[37] This is Klein's paradox.*[31]

Free neutrons are unstable, but most atomic neutrons are stable. According to the quantum mechanical model of the nucleus, the protons and neutrons are arranged according to the specific energy levels with quantum numbers in the nucleus. In order for a neutron to decay, the proton must be in a state of lower energy than that of the neutron. In a stable nucleus, this is quantum-mechanically impossible because all energy levels are filled and no lower energy state exists. Unstable neutrons in atoms are lost by beta decay. For example, $^{14}C(6p + 8n)$ decays to ^{14}N $(7p + 7n)$ (half-life 5,730 years). Within the nucleus, a proton can turn into a neutron if there is a quantum mechanically possible energy state. In this case, a positron and a neutrino are emitted, and the produced positron is immediately combined with an electron and eliminated.[38]

---

*[31] In 1929 Oskar Klein (9.1894. - 2.1977, Swedish theoretical physicist) found a surprising fact when applying the Dirac equation to the problem of electron scattering by a potential energy wall. In non-relativistic quantum mechanics, electrons decay exponentially as they pass through this wall. However, Klein's results showed that if the potential energy is about the electron mass, the wall is almost transparent. Moreover, when the potential energy goes to infinity, the reflection decreases and electrons always pass through. This contradiction was applied to Rutherford's proton-electron model for the neutral particles in the nucleus before neutrons were yet discovered. This contradiction eliminated the concept that electrons are inherent in the nucleus. This apparent contradiction means that electrons cannot be trapped within the nucleus by any potential energy.

2.2. History of atomic model

As is well known, ordinary matter consists of atoms. Atoms are not elementary particles that make up ordinary matter, but they determine most of the physical properties of materials. The atom is known to consist of sub-atomic particles such as protons, neutrons and electrons. In particle physics, the electron is an elementary particle that cannot be divided into smaller ones, and the proton and neutron are composed of elementary particles called quarks. Quarks are particles that can only be observed indirectly in very specific environments and need not be discussed in solid-state physics. As far as the electromagnetic properties of solids are concerned, there is no room for quarks to do something. In solid-state physics, the physical properties of solids are determined by how their constituent particles, the electrons, protons, and neutrons are arranged, and how these atoms combine or interact with other atoms. This of course depends on temperature and pressure. It is stated here how the atoms that make up solids are constructed and whether there are any loopholes in these current atomic models, and whether there is any possibility of other atomic models.

Thomson's atomic model

The etymology of "atom" is known to be "atomos", an ancient Greek adjective meaning "undivided".[39] In the twentieth century, numerous physics experiments showed that the "undivided" atom actually consisted of electrons,

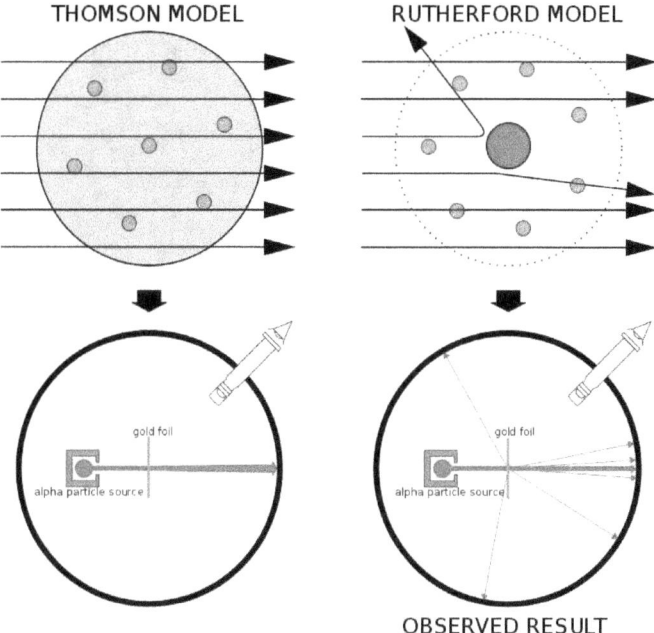

Figure 9. Geiger-Marsden Experiment (image from wikipedia.org): (left) Prediction by the Thomson model, if alpha particles pass through the atoms of the plum pudding model, there is no deflection. (right) Experiment, some alpha particles were deflected (by positive charges concentrated in the nucleus).

protons, and neutrons. J.J. Thompson confirmed in 1897 the presence of electrons in th cathode ray experiments using a device such as shown in Figure 4.[40] When an electric field is applied between the two left electrodes of a sealed glass vessel maintained in a vacuum, a beam is generated at the cathode and hits the glass surface opposite the glass tube. If another electric field is applied to the electrodes in the center of the glass vessel, the beam is deflected by the electric field. The beam was not of light, but of a very light stream of negative particles, and the mass-to-charge ratio indicated that the mass was about 1,800th of a hydrogen atom. Thus, it was assumed that there are smaller building blocks than the atoms.[41] In those days when the existence of nucleons, protons and neutrons was unknown, Thompson considered these negative particles to be negative raisins embedded in the positive charge pudding and proposed the so-called plum pudding model (upper left in Figure 9).[42]

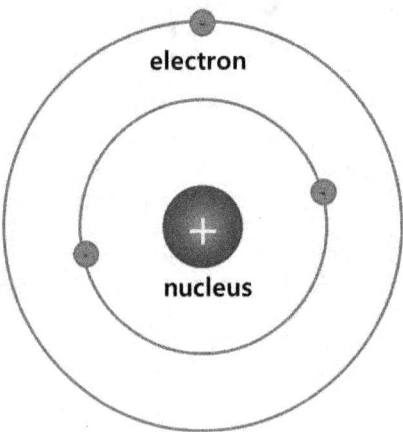

Figure 10. Bohr's atomic model.

Discovery of atomic nuclei and new atomic models

However, Thompson's atomic model was denied by Rutherford in 1909. Rutherford, together with Geiger*[32] and Marsden*[33] in 1909, performed an experiment called the

---

*32) Johannes W. Geiger (9.1882 - 9.1945) was a German physicist. Known as the co-inventor of Geiger counters.

Geiger-Marsden experiment in which alpha particles (helium nuclei) were shot on a thin metal plate to measure deflections on a fluorescent screen.[43] In the Thomson model, the positive charges spread like a pudding, and because of their low density and the very low electron mass compared to that of alpha particle, all the alpha particles were expected to pass through the metal film without deflection. However, very rarely, some alpha particles deflected at large angles, as shown in Figure 9 (the upper and right). Rutherford deduced that the positive charges of the atom were confined to a very small volume, which had led to the deflection of alpha particles strong enough. Most the atomic masse and positive charge were found to be concentrated in very narrow zones.

Rutherford proposed a new atomic model in which the electrons orbit like satellites around the central nucleus.[44] However, this atomic model had two serious problems. Unlike the planets orbiting the Sun, the electrons are charged particles. Rotating in an orbital motion emits electromagnetic waves, and when the energy is lost, it will be attracted to the nucleus and crash into it instantly. This is the first problem, and the second one is that the atomic model does not explain the absorption and emission of electromagnetic waves by atoms in a very narrow wavelength range.

During the early 20th century, and the early years of quantum mechanics, a hypothesis was suggested that the energy absorbed and released by the electrons is quantized. In 1913 Bohr*[34] proposed an atomic model, the Bohr model

---

*[33] Sir Ernest Marsden (2.1889 - 12.1970) was a British-New Zealand physicist.

(Figure 10), in which the electron orbits around the nucleus with a fixed angular momentum and energy. The distance to the nucleus is proportional to the energy. In this model, the electron does not lose energy continuously and is not absorbed by the nucleus. Only it loses energy via momentary "quantum jumps", when an electromagnetic wave having the frequency equal to the difference in energy is absorbed or emitted.[45] This model can explain the simple spectrum of hydrogen, but it does not apply to the atoms with many electrons. Owing to the development of spectrometry, additional absorption lines (see Figure 17) were found in hydrogen that were not explained by the Bohr model. In 1916, Sommerfeld refined the Bohr model by adding an elliptical orbit, but no other atoms have yet been well described.

In 1924 Schrödinger,[*35)] inspired by de Broglie matter wave, thought that electrons would be better explained as waves than particles. The Schrödinger equation,[*36)] published in 1926, describes electrons in terms

---

[*34)] Niels H.D. Bohr (10.1885 – 11.1962) was a Danish physicist. He was awarded the Nobel Prize in Physics in 1922 for his contribution to the understanding of quantum theory and the atomic structure. Bohr introduced the complementarity principle that any system can be interpreted as a wave or a particle. This means that a quantum mechanical object can behave as a particle or a wave in some aspects. It is called "complementary" in that it cannot be a particle and a wave simultaneously.

[*35)] Erwin Schrödinger (8.1887 – 1.1961) was an Austrian physicist. He was awarded the Nobel Prize in Physics in 1933 for the Schrödinger wave equation. He also made many achievements in statistical mechanics, thermodynamics, electromagnetism and general relativity.

[*36)] Schrödinger equation is a linear partial differential equation

of wave function, rather than as discrete particles.[46] This equation well explained the spectroscopic phenomena of atoms. It was mathematically convenient but not easy to understand. The wave equation does not describe the electron itself, rather describes its possible states and can be used to calculate the probability of finding an electron at a specific location around the nucleus.[47]

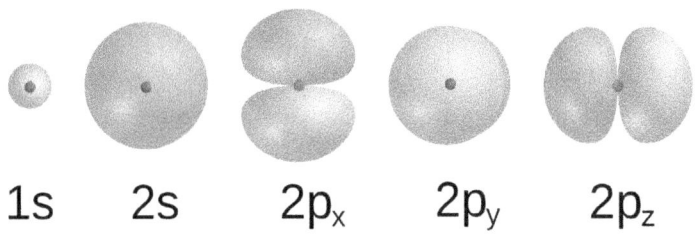

Figure 11. Five energy levels of neon (image from wikipedia.org). Each energy level can have two electrons.

Describing an electron in terms of wave function makes it impossible to mathematically derive the electron's position and momentum at the same time. This is Heisenberg's uncertainty principle and was published in 1927.[48] This obsoleted the Bohr model. The electron that makes up an atom is represented by the probability of being found at a particular location. This probability depends on the energy level and is called the atomic orbital. Figure 11 shows the energy levels of neon.

---

that describes the evolution of non-relativistic quantum mechanical systems over time. This is the basic equation of wave mechanics invented by Schrödinger.

2.3. Hydrogen and helium

Hydrogen occupies 75% of ordinary matter in the universe in the form of a single atom. Most are protium (^1H) free of neutron. In the standard model of cosmology, it was first generated during the recombination epoch when electrons and protons were separated after the Big Bang.[49] Hydrogen forms covalent bonds easily with nonmetallic elements. The ground energy of hydrogen is -13.6 eV, which matches with an electromagnetic wave of ~91 nm in the ultraviolet region. When a hydrogen atom in the ground state absorbs a light wave of this wavelength, it decomposes into an electron and a proton.

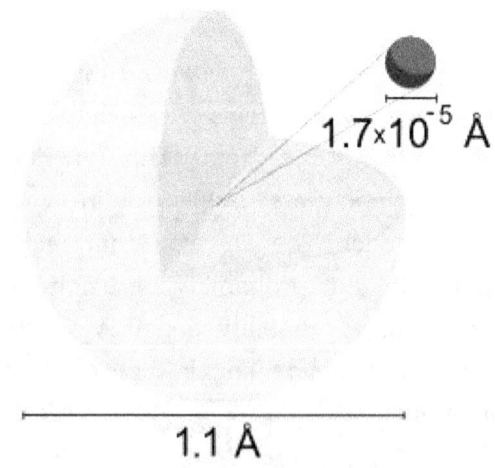

Figure 12. Hydrogen atom model (image from wikipedia.org).

Hydrogen - a proton-electron pair

The previous hydrogen atom models depicts an electron to orbit around a proton. A typical example is the Bohr model shown in Figure 10. In this atomic model, the protons are in the center, and the electrons orbit around them like the solar planets around the Sun, but the interaction is not gravity but electromagnetism. Advances in quantum mechanics have led to several theories (Schrödinger equation, Dirac equation,[37] or the Feynman[38] path integral).[39] Calculations of the electron probability density

---

[37] It was published by Paul Dirac in 1928. It is a relativistic quantum wave equation describing fermions with a spin of ½. The Dirac equation follows the mirror symmetry and space-time reversal. For this reason, this equation is used when describing electrons in the theory of mirror symmetry (quantum electrodynamics, etc.).

[38] Richard P. Feynman (5.1918 - 2.1988) was an American theoretical physicist. He is well known for Feynman's path integral in quantum mechanics, a theory of quantum electrodynamics. He also contributed in the physics of superfluidity for supercooled liquid helium, as well as in particle physics. For his contributions to quantum electrodynamics, he received the Nobel Prize in Physics in 1965.

[39] In quantum mechanics, the path integral is a way of describing quantum theory by generalizing Hamilton's principle. The probability of transiting from one state to another is the functional integral of all possible paths between the two states. It was firstly introduced by Dirac and in 1948 it was further developed into a specific methodology by Feynman. This method has a tremendous effect on theoretical physics because of the symmetrical description of time and space. In classical mechanics, the motion of a point particle or mass point can be determined entirely from the initial state by solving the equation of motion. Because of uncertainty in quantum mechanics, one cannot think of just one path as in classical

around the proton using these theories shows no angular momentum. It means that the electron does not orbit the proton.[50] These results can be expressed by the hydrogen atom model shown in Figure 12. It is similar to the concept of the new vacuum paradigm in the sense that the probability distribution is interpreted as a strain field of the solid vacuum surrounding the proton. The strain energy is the hydrogen atom energy.

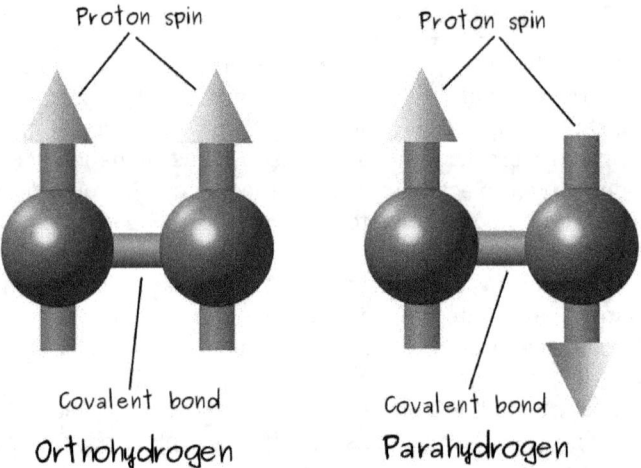

Figure 13. Spin isomers of molecular hydrogen (image from wikipedia.org).

---

mechanics. In the path integral, quantum mechanical probabilities are obtained by including all the infinitely many paths connecting the starting and ending points in space-time. Schrödinger wave mechanics and Heisenberg's matrix mechanics deal with problems with the equations of motion, while the path integral deals with quantum mechanics over the entire paths, focusing on the path of motion.

Hydrogen molecules are classified according to the spin state of the nucleus.[51] Ortho-hydrogen of the triplet state where the proton spins of the two atoms are parallel and thus the total spin of the hydrogen molecule is 1 (= ½ + ½), and para-hydrogen of the singlet state where the spins are opposite and the total spin is 0 (= ½-½). At the standard condition of temperature and pressure, 25% of hydrogen molecules are para-hydrogen and 75% are ortho-hydrogen.[52]

Ortho-hydrogen is an excited state and at cryogenic temperatures almost all the hydrogen molecules are para-hydrogen. The thermal properties of liquid and vapor phases are very different due to the difference in the rotational heat capacity.[53] The change from ortho- to para-hydrogen is exothermic and generates enough heat to evaporate some liquid hydrogen.

Most of hydrogen (99.98%) is ^{1}H with a proton as shown in Figure 12. Deuterium (^{2}H) consisted of a proton and a neutron in the nucleus, and tritium (^{3}H) with a proton and two neutrons are also found in the nature. ^{2}H is stable, but ^{3}H is a radioactive element with a half-life of 12.32 years. It transforms to helium-3 (^{3}He, 1 neutron + 2 protons) through beta decay in which the highly energetic neutron becomes a proton.[54]

Helium – the most stable atom bound by two neutrons

Helium (He) with the atomic number 2 is the second most abundant element in the universe after hydrogen, occupying about 24% of the total matter. The amount is similar to that of the Sun. Most of helium in the universe is helium-4 (^{4}He, 2 neutrons + 2 protons). According to the Big Bang theory, it is thought to have formed during the Big Bang. A significant amount of helium is produced by nuclear fusion of hydrogen. It can be intuitively inferred that ^{3}H produced during the hydrogen fusion process becomes ^{3}He after beta decay and eventually turns into more stable ^{4}He. Hydrogen fusion requires tremendously high pressures and temperatures of over 100 million degrees Celsius. The interiors of a star, like the Sun, are highly stressed, energy-concentrated by the surrounding solid vacuum,*40) and continuously release energy through fusion. The most fundamental process is hydrogen fusion, represented by the proton-proton chain reaction shown in Figure 14. In this process, energy is released in the form of gamma rays and neutrinos. This reaction occurs mainly in the Sun or smaller stars. Meanwhile, a catalytic cycle, another type of fusion reaction called the CNO cycle,*41) occurs in stars 1.3 times more massive than the Sun.[55] Proton-proton fusion reactions occur in general when they have enough energy to overcome the Coulomb repulsion between them.[56] The production of deuterium is rare in the Sun. This is because the proton pair generated by proton-proton fusion immediately separates into two

---

*40) Please refer to Chapter 3 of "ORIGIN OF GRAVITY & NEW COSMOS" ISBN 9781713042020. In the regime of the new vacuum paradigm, mass is nothing but the vibration energy of the solid vacuum, which induces stresses in the surrounding solid vacuum by distorting it as much as the energy of mass.

*41) A nuclear fusion reaction in which Carbon-Nitrogen-Oxygen are involved in the production of helium from hydrogen. In the CNO cycle, four protons are involved from carbon, nitrogen and oxygen isotopes to produce one alpha particle and two positrons and two electron neutrinos.

protons. Due to the slow transition of hydrogen to helium, it is predicted that the transition from hydrogen into helium in the solar core will take more than 10 billion years.57

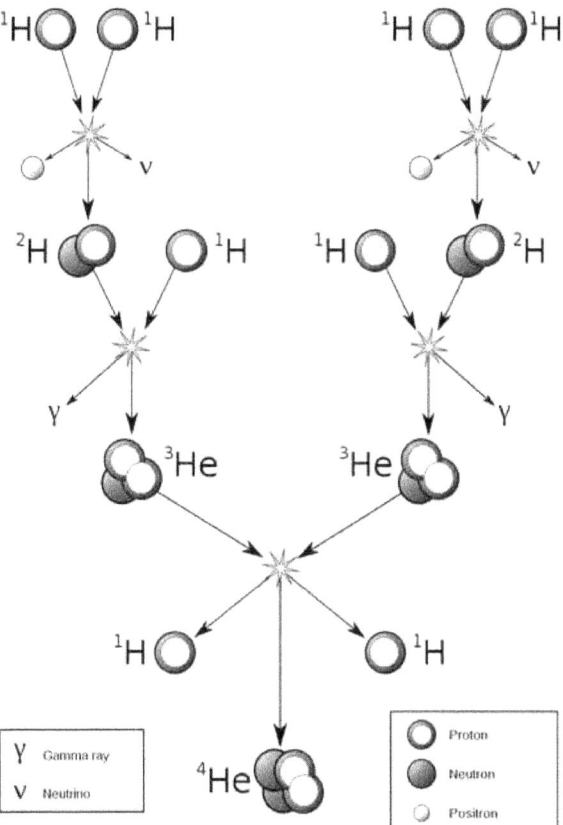

Figure 14. The proton-proton chain reaction in the solar core (modified image from wikepedia.org).

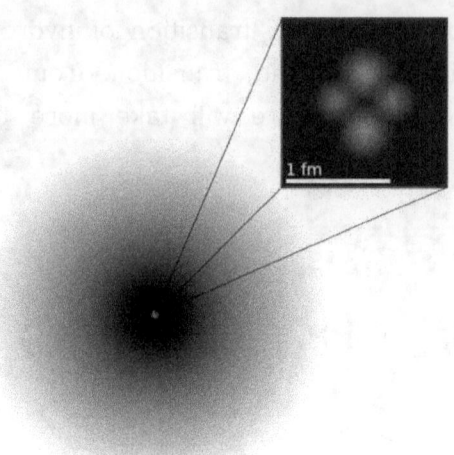

Figure 15. Helium atom. Nucleons have a spherical symmetry (image from wikipedia.org).

Classically, the helium atom is imagined to have two electrons orbiting around the nucleus made of two protons and two neutrons. Quantum mechanically, as with the hydrogen atom, the electrons do not rotate around the nucleus and the movement is replaced by the electron probability density around the nucleus and there will be no net angular momentum. Since helium has more nucleons than hydrogen, its energy is more concentrated (some energy is released as gamma rays or neutrinos) and therefore will exert greater pressure to the solid vacuum.

The nucleus of ^{4}He in Figure 15 is the same as an alpha particle in composition. In high-energy electron scattering experiments, the charge of helium decreases exponentially away from the charge center. This symmetry reflects that the proton pair and the neutron pair follow the same law of quantum mechanics in the helium nucleus as the electron pair follows quantum mechanics. That is, all the nucleons occupy 1s orbitals, and have no angular momentum and cancel each other's spins. This configuration is very unusually stable energetically. This is why helium is a very stable element in the nature.*42)

^{3}He, on the other hand, is composed of two protons and a neutron and is the only stable element among the elements with the smaller number of protons than neutrons. It is only 0.02% of ^{4}He in the amount. One neutron is deficient and thus is lighter than ^{4}He and its ground state energy is relatively high, so its physical properties are slightly different. Lower thermal energy input than with ^{4}He breaks the bond formed by the dipole-dipole interaction. ^{4}He is classified as boson because the total spin is zero, but the sin of ^{3}He is ½, hence it is a fermion. ^{3}He boils at 3.19 K and ^{4}He boils at 4.23 K. The latent heat of vaporization of ^{3}He is 26 J/mol while it is 82.9 J/mol for ^{4}He.58 Therefore, although ^{3}He is a stable element, it is energetically less stable than ^{4}He.

2.4. New hydrogen atom model

When (vibration) energy is stored in the vacuum lattice, the lattice vibration distorts the solid vacuum around it. For example, the existing atomic model for hydrogen secures

---

*42) This argument is from helium@wikipedia.org. But no reference can be found in there.

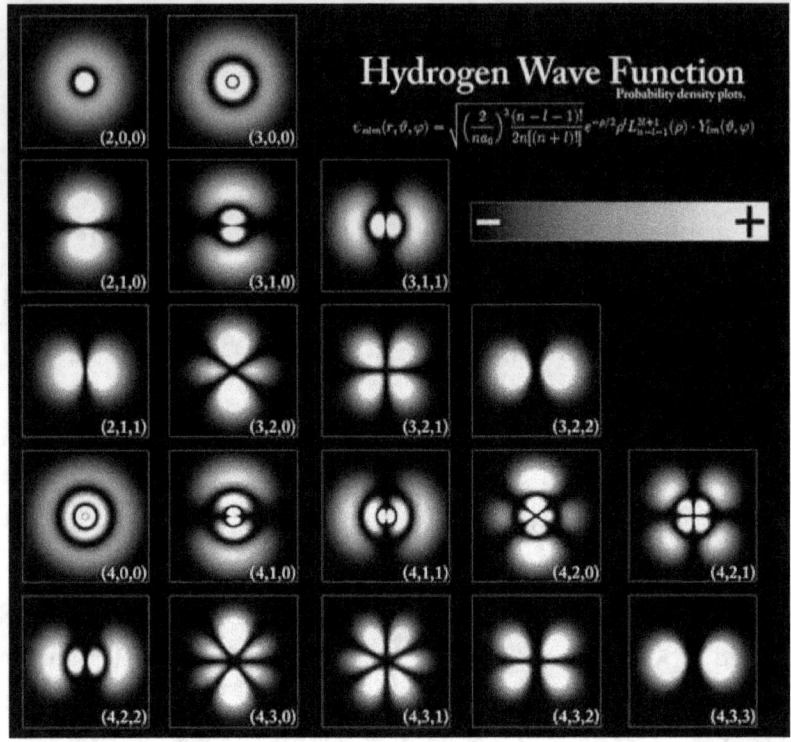

Figure 16. Cross-sections of the electron orbitals in hydrogen (image from wikipedia.org).

an atomic volume as the electron vibrates around the proton. It is similar to the concept that a circular electromagnetic wave, as shown in Figure 3, surround the proton. Recent studies confirm no electron angular momentum for the atomic electrons. The atomic model in the new paradigm ignores the presence of electrons as discrete particles. In this model the elementary lattice point of the solid vacuum expand to form a stress field inversely proportional to the square of the distance from the center of expansion. Therefore, it is assumed that the proton and electron in the hydrogen atom of Figure 15 are not distinguished separately but are indicative of the stress field developed in the solid vacuum due to the volume expansion via vibration. As more energy is accumulated in there, the volume will further increase and simultaneously the surrounding solid vacuum will be distorted as much as the volume expansion. This expansion does not occur homogeneously (or continuously) but anisotropically with an increase in energy. This is because the solid vacuum forms a regular lattice, and in order to increase its volume, it must interact with the lattice structure of the solid vacuum. This process can be inferred from the hydrogen electron orbital cross sections in Figure 16 and the absorption spectrum of hydrogen in Figure 17.

Circular standing waves - electron and positron

In the new paradigm, the electron is a circular standing electromagnetic wave, as shown in Figure 3. In this standing wave, the periodic change (vibration) of the electric field (or magnetic field) propagates into the solid vacuum in a point symmetrical mode, and this propagation generates a force inversely proportional to the distance, the electromagnetic force. When a linear electromagnetic wave becomes circular, the state of distortion of the inner and outer electric fields of the circular wave is different, and the particles outside the circular electromagnetic wave feel a positive or negative electric field depending on the arrangement of the standing wave. When a negative electric field is felt, we have an electron, and when a positive electric field is felt, we have a positron. Around this circular electromagnetic wave, the solid vacuum is distorted and stressed and an indued stress develops in the direction perpendicular to the rotational plane of the standing wave. This is the magnetic moment of electron or positron. From an electromagnetic wave, an electron-positron pair can be formed, and the positron is captured in a vacuum lattice and becomes a proton. When the proton and electron fuse, they become a neutron. If the electron is a standing electromagnetic wave, it can be imagined that the absorption of light increases the frequency of standing waves of the electron, while the distance from the center increases when neutrinos are absorbed or emitted.

New hydrogen atom model

In the new atomic model, a hydrogen atom is a vibrating lattice point of the solid vacuum by absorbing energy (a lattice point becomes a proton by capturing a positron and an electron is trapped around it). This vibration emits matter wave and exerts a compressive stress to the surrounding solid vacuum, which is called the probability distribution of electron in quantum mechanics. The stress field in the solid vacuum appears to be a volume expansion of the solid vacuum. Hence, the volume of a hydrogen atom can be regarded as the stress field formed by the electron trapped in the stress field of the proton. In our previous work, Origin of Gravity and New Cosmos, gravity is originated from the vibration called mass, the force of which is inversely proportional to the square of the distance from the center as the vibration propagates as matter wave radially. Thus, the force of this stress field allegedly due to the electron distribution is naturally inversely proportional to the square of the distance from the center. This is the electromagnetic force known as the force between the two opposite electric charges.

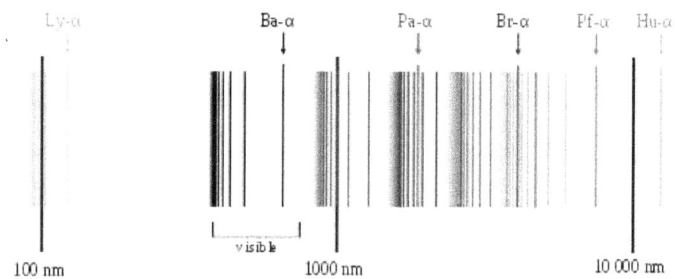

Figure 17. Hydrogen absorption spectrum(image from wikipedia.org).

Individual hydrogen atoms become energetically more stable when they form molecules. As mentioned previously, there are two types of hydrogen molecules, ortho- and para-hydrogen depending on the molecular spin. When one of the atoms of a hydrogen molecule in an environment as in the solar core releases energy and turns into a neutron by merging its proton-electron pair, the hydrogen molecular becomes a deuterium. In the deuterium atom of the new atomic model, the proton is not together with the neutron, but is together with the electron. Namely, a deuterium atom is imagined to have the composition of a hydrogen atom and a neutron and further imagined to be in dynamic equilibrium between the hydrogen atom and the neutron in the deuterium atom, Namely, when the hydrogen atom becomes a neutron, the existing neutron become a new hydrogen, as

$$p^+ + e^- \Leftrightarrow n^0 \quad (2.1).$$

In this way, the deuterium atom can mitigate the asymmetric surrounding stress field. When heavy elements are synthesized in the stellar or solar core with very high pressures, hydrogen atoms will transform to hydrogen molecules, deuterium, tritium, and helium progressively, to increase the number of nucleons in the atom (Figure 14 is another representation for this process). When the number of heavy elements increases and the star explodes at the critical point (due to the rapid increase in the internal energy via nuclear fusion), a supernova is born, and the heavy elements are spreading throughout the universe, where they may recombine to form an Earth-like planet. The various elements of Earth come from supernovae and the less energetic heavy elements are centered because of the high pressure in the Earth core.*[43]

2.5. Atomic structure in new atomic model

Various elements we know can be summarized in the periodic table as shown in Table 1. Why are there more than 100 kinds of element? Where do these elements come from? More than 75% of matter in the universe is hydrogen

---

*[43] Refer to section 3.5. Supernova of "Origin of Gravity and New Cosmos"(Min Tae Kim. ISBN 9781713042020)

Table 1. Periodic Table of Elements

원소 주기율표
Periodic Table of Elements

H																	He
Li	Be											B	C	N	O	F	Ne
Na	Mg											Al	Si	P	S	Cl	Ar
K	Ca	Sc	Ti	V	Cr	Mn	Fe	Co	Ni	Cu	Zn	Ga	Ge	As	Se	Br	Kr
Rb	Sr	Y	Zr	Nb	Mo	Tc	Ru	Rh	Pd	Ag	Cd	In	Sn	Sb	Te	I	Xe
Cs	Ba	La-Lu	Hf	Ta	W	Re	Os	Ir	Pt	Au	Hg	Tl	Pb	Bi	Po	At	Rn
Fr	Ra	Ac-Lr	Rf	Ha													

La	Ce	Pr	Nd	Pm	Sm	Eu	Gd	Tb	Dy	Ho	Er	Tm	Yb	Lu
Ac	Th	Pa	U	Np	Pu	Am	Cm	Bk	Cf	Es	Fm	Md	No	Lr

and the rest is almost helium. Hydrogen is the main component of stars. In other words, a star is agglomerated hydrogen atoms. In the regime of the new vacuum paradigm, a hydrogen atom is a deformed structure of the solid vacuum where its energy is stored in the form of paired proton and electron. The surrounding solid vacuum is also distorted, which in turn affects the hydrogen atom to be subjected to a compressive stress. This compressive stress and the nature of matter wave cause matter including hydrogen atoms to coalesce.*44) The aggregated hydrogen atoms become a spherical star in order to reduce the internal energy. As it gets to the stellar core, the stress increases even to a singularity at the very central point. Under these very high stresses, hydrogen molecules or atoms are compressed and separated electrons and protons fuse into neutrons and release energy to the outside. As this process repeats, heavy elements are created as

---

*44) It is conventionally called gravity.

hydrogen atoms get more neutrons as the binders. In stars like the Sun the amount of heavy elements other than helium are insignificant. On the contrary, Earth has a very large amount of heavy elements such as iron compared to the amount of hydrogen and helium. Earth's constituents should have been produced via nuclear fusion reactions in the stellar cores for a long time. When a supernova explodes, the heavy elements produced in the progenitor star spread out into the space, and when the high heavy elements get together, they may form Earth-like planets.

Periodic variation of elements - The periodic table

According to existing physics theory, the fact that the elements known to us are categorized into a periodic table shown in Table 1 is due to the electron distribution around the atomic nuclei. In conventional atomic models, the protons and neutrons are gathered at the atom center, and the electrons are arranged at the outer regions according to the four indices representing the electron energy levels. According to the Pauli*45) exclusion principle,*46) two electrons in an atom cannot have the same value of the four indices.*47) This atomic model would be highly

---

*45) Wolfgang Ernst Pauli (4.1900 - 12.1958) was an Swiss and American theoretical physicist. One of the pioneers of quantum mechanics. In 1945 he won the Nobel Prize in Physics for discovering a new law of nature called the Pauli exclusion principle.

*46) It is a principle of quantum mechanics discovered by Pauli in 1924 stating that two identical fermions (particles with a ½ spin like electrons) cannot coexist in a quantum state. According to this principle, two electrons in an atom cannot have the same quantum number.

plausible for light elements, such as hydrogen or helium. But such a configuration is hard to imagine for heavy elements, for example, uranium (U), because U should have 92 electrons around its 238 nucleons. The new paradigm assumes that the electron and proton are always paired in the atom.*48) One proton has one electron like a hydrogen atom (we call this electron-proton pair a "plecton". Its composition is the same as ^1H, but it belongs to heavier elements than hydrogen). When an atom absorbs energy, it is stored in the form of the potential energy of the plectons in the atom. This energy is stored as the distortion energy of the solid vacuum. The generation of atoms in the solid vacuum can be thought of as the inserting foreign bodies of different sizes into a uniform (crystal) structure. If a foreign body is present, the solid vacuum will be deformed and the mode of deformation will be dependent on the structure of the solid vacuum. In this sense, the electron configuration in the atom should be related to the structure of the solid vacuum.

Grape bunch atomic model

What is the atomic structure in the new paradigm? If a hydrogen molecule consisted of two hydrogen atoms releases the energy of one hydrogen atom by nuclear fusion, it will become deuterium (1 plecton + 1 neutron).

---

*47) n (the principal quantum number), l (the angular momentum quantum number), m_l (the magnetic quantum number), and m_s (the spin quantum number.)

*48) This assumption is well suited to the Pauli exclusion principle. This is because more than one electron cannot be paired with a proton in the atom.

We assume this deuterium will be a elementary subatomic component in the new atomic model. In a deuterium atom the electron (e^-) is assumed to freely absorbed to the proton (p^+) and ejected from the neutron (n^0) and vice versa in dynamic equilibrium as in Eq. (2.1).

The forward and reverse reactions of Eq. (2.1) will repeat constantly in the interiors of a Sun-like star. Deuterium is a stable element because the process of Eq. (2.1) can occur within the atom (when the plecton turns into a neutron, while the other neutron turns into a plecton, the deuterium atom continues to exist). In fact, deuterium is a very stable element, but its extinction rate in the star higher than its rate of formation.*49) Deuterium in the universe is estimated to be 0.02% of hydrogen, which is probably due to the conversion of deuterium to helium. The fact that helium accounts for 24% of ordinary matter in the whole universe supports this hypothesis. A helium atom is composed of two deuterium atoms and has higher symmetry than one deuterium atom. If the reaction of Eq. (2.1) takes place alternately within one helium atom, there will be little energy exchange with the surroundings. This is why helium is so stable. Therefore, it is natural for the helium structure to be the unit structure of heavier elements. We propose a new atomic model in which plectons are stacked to form multiples of the helium structure in a spherical form with the minimum volume for heavier elements.

In order to build an atom in a spherical form, the number of protons (namely, plecton-neutron pairs, the building blocks) should increase in proportion to the

---

*49) It is mentioned in "Deuterium" of wikipedia.org, but there is no reference so there is no clear evidence.

Table 2. Atomic arrangement according to the spherical atomic structure (r = radius)

r	r^2	No. of proton	element	subatomic configuration	period
1	1	1×2=2	He	$2p+2n$	1
2	4	4×2=8	Ne	[He]+$8p+8n$	2
2	4	4×2=8	Ar	[Ne]+$8p+8n$	3
3	9	9×2=18	Kr	[Ar]+$18p+18n$	4
3	9	9×2=18	Xe	[Kr]+$18p+18n$	5
4	16	16×2=32	Rn	[Xe]+$32p+32n$	6
4	16	16×2=32	Og	[Rn+$32p+32n$	7
5	25	25×2=50	-	[Og]+$50p+50n$	8
5	25	25×2=50	-	[?]+$50p+50n$	9

surface area, which is proportional to the square of the radius. Therefore, the subatomic arrangement of plecton-neutron pairs will be given as shown in Table 2, where r is the dimensionless radius. At $r=1$, the number of protons is 2, namely 1×2=2, which gives helium. At $r=2$, we have 2^2×2=8 for neon. Currently, oganesson*[50]) in the $r=4$ group is known to be the heaviest,[59] and elements in the $r \geqq 5$ group have not been found or synthesized yet. In this table, two shells exist at the same r value. What shall be the reason for that?

---

*[50]) Og, the atomic number 118, was artificially synthesized in 2002 at the Joint Institute for Nuclear Research in Russia for the first time in a joint study by Russian-American scientists, It was named after Russian nuclear physicist Yuri Tsolakovich Oganessian (4.1933 -) in 2016.

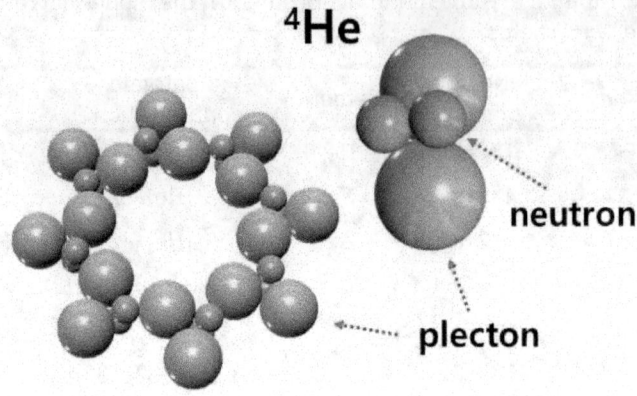

Figure 18. An imaginary two-dimensional outmost shell in an atom consisting only of helium blocks

There is alpha decay, a kind of radioactive decay. The process emits helium nuclei, or alpha particles. As introduced in Section 2.3, helium is a very stable element and the atomic number increases as the helium structure is stacked. If the primary shell composed of PeNs[*51)] (1 PeN = $1p1e + 1n$) is formed for a given r, the secondary shell will complete the helium structure by resting on the primary shell. For example, neon has the primary shell composed of 8 PeNs at $r=2$, the secondary but the same shell of additional 8 PeNs gives argon. That is, two shells are required at a given radius of r to have the complete spherical helium structure. Figure 18 shows a two-

---

[*51)] PeN is a compound word using Proton, electron and Neutron's first alphabets. It is the elementary building block of atoms in the new paradigm and one PeN has the same composition with a deuterium atom. It is the triplet mentioned in the preface.

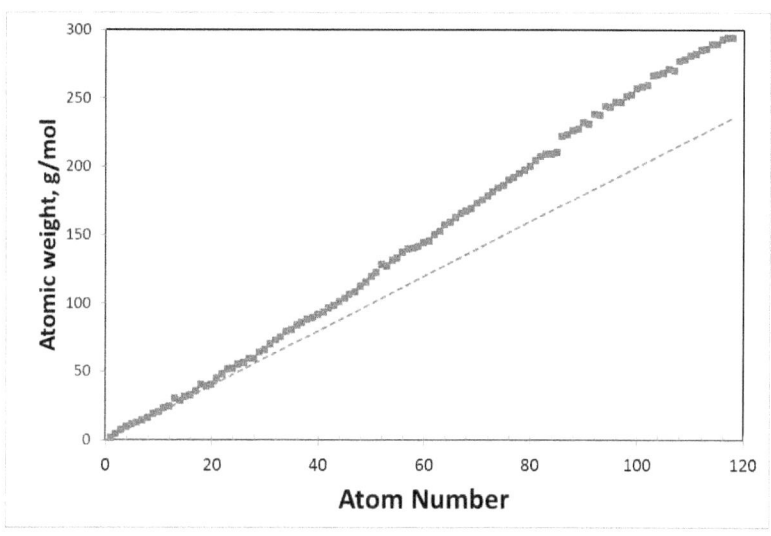

Figure 19. Atomic number vs. atomic weight. Dotted line indicates (atomic number)×2.

dimensional schematic of an atom whose outermost layers all consist of helium blocks.

If a PeN is the primary building block of atoms, then exactly twice the atomic number is its atomic weight. However, as shown in Figure 19, there is a difference between twice the atomic number and the actual atomic weight. This difference increases with the atomic number. The difference is called the neutron excess. Why does the neutron excess increase with the atomic number? In the regime of the new vacuum paradigm, as the atom grows in size (higher in the atomic weight), the stress at the atomic core due to the surrounding solid vacuum will also increase. Accordingly, it will be difficult for energetic plectons to exist within the atomic core because the stress will increase exponentially on going to the core. Thus, to

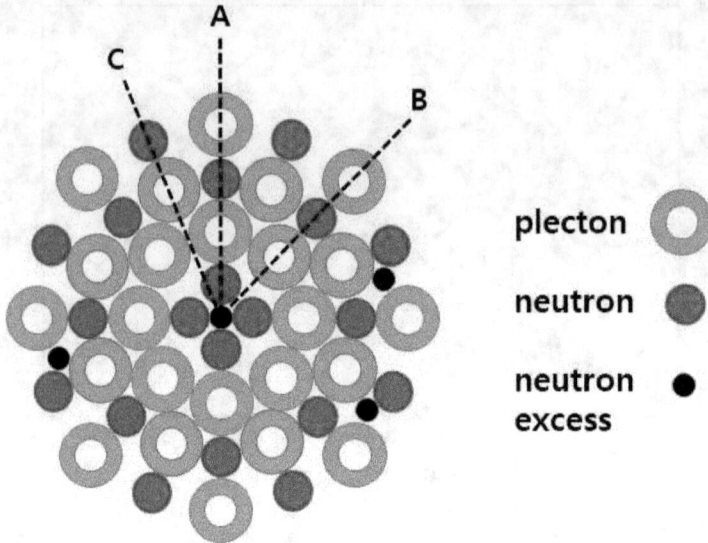

Figure 20. An imaginary two-dimensional heavy atom

form a stable atom, the core would have to be filled with less energetic neutrons. Imagine an atomic structure with less energetic excess neutrons in the center, as shown in Figure 20.[*52] The heavier the element, the more excess neutrons are present, being concentrated in the atomic center. These excess neutrons are regarded as the neutrons run out of energy in the new vacuum paradigm. Hence the heavier is an atom, and the more stable energetically. In this context, a large neutron star is an atom with a great number of excess neutrons inside it, and thus with a very large atomic number.[*53]

[*52] Only one excess neutron is shown in the figure. It can also be imagined that the excess neutrons are dispersed throughout the atomic structure as shown.

[*53] It is estimated that there will be elements of the atomic

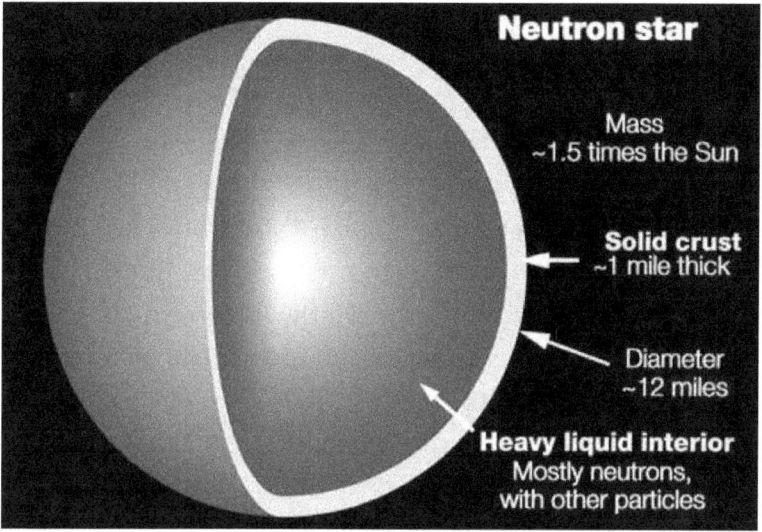

Figure 21. The structure of a neutron star (image from wikipedia.org). A neutron star is a very heavy atom with a high density of excess neutrons in its large core.

Atoms synthesized under very high stresses in the interiors of massive stars are in the process of continually being compressed, so that their constituent plectons (equivalent to ^1H) turn into neutrons and heavier elements are synthesized. When these elements are exposed to free space via supernova explosion (or Earth's internal matter is exposed to the atmosphere), and thus the stress is removed, it is inevitable that the energetic (new born) neutrons decompose into protons and electrons in the reverse reaction of fusion (beta decay) or are emitted to be free neutrons. This process is called nuclear fission, where heavier elements are converted to lighter ones,

---

number over 119 in the universe that we did not find yet.

accompanied by the emission of nuclear energy. Because the reaction is suppressed in the stellar core due to high internal pressures, the unstable radioisotopes of heavy elements releases energy without emitting neutrons, turning into stable ones.

If the solid vacuum is tremendously hard, not comparable to ordinary matter,*54) and if the lattice constituting the solid vacuum are even very slightly deformed due to vibration, the magnitude of stress in the surrounding solid vacuum can be very large. In this sense, the electron is a stress field formed around the virtual vacuum lattice (the electron itself is known as having zero volume). The probability distribution of the electron in the hydrogen atom model is regarded to be the magnitude of this stress field. An element is a thing composed of these stress fields, and the electronic energy level of an atom is a representation of the stress field associated with the structure of the solid vacuum. This will be similar to the stress field formed around impurities, vacancies and dislocations in metallic crystal lattices. This is discussed in the next chapter.

---

*54) From the propagation speed of the shear electromagnetic wave called light, it can be assumed that the modulus of elasticity is very high.

III. Chemical bond

All ordinary matter is nothing but the vibration of vacuum lattice. Mass is just the frequency of the vibration and distorts the solid vacuum by emitting matter wave. The intensity of this distortion is inversely proportional to the square of the distance from the center of vibration. Things in the neighborhood are attracted towards the more severely distorted solid vacuum caused by matter wave. Things get together and the whole energy is reduced as they get along. Atoms in the periodic table form bonds to reduce their total energy. Each atom occupies the optimal location in materials to reduce the energy accumulated in the surrounding solid vacuum. This is the chemical bond. In chemical bonds, some bonds form between the same kind of atoms and some bonds forms between the different kinds of atoms. In addition to these primary bonds, secondary interatomic bonds are sometimes formed in solids and in massive molecules. The nature of these bonds depends on how the triplets, namely PeNs that make up an atom are arranged. It is mainly based on the number and location of valence PeNs in the outermost shell. Modern physics calls this the electron configuration showing the electron energy level in an atom. In this chapter, we will analyze the electronic energy level based on the new atomic model and re-interpret the nature of various chemical bonds.

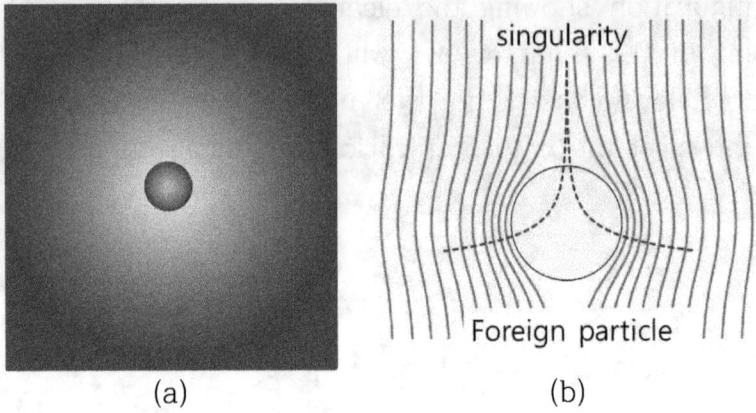

Figure 22. Distortion of the solid vacuum around a star (a) and a stress or strain field developed around a bulky atom inserted in a homogeneous solid (b).

3.1. Energy level and stress field around crystal defects

Stress field around crystal defects

The presence of vacancies*[55] or particles of other kinds in a crystal, such as a pure solid metal, results in local distortions and stress fields.[60] A spherical vacancy or an impurity in a very homogeneous material induces a radially symmetrical stress field, as shown in Figure 22. This is similar to the case with the Sun or Earth in the solid vacuum. On the other hand, if there is a dislocation in a

---

*[55] In crystallography, vacancies refer to point defects in a crystal. If an atom in a lattice point is empty in the crystal, this empty lattice point is called a vacancy.

crystal, the surrounding stress develops intricately. A dislocation is a displacement by the size "Burger's vector, b" shown in Figure 23 from the normal position in a regular crystal structure, as if one button was inserted incorrectly to impair the overall clothing balance.*56) The stress field around a dislocation accumulates up to dozens of eV of energy.61

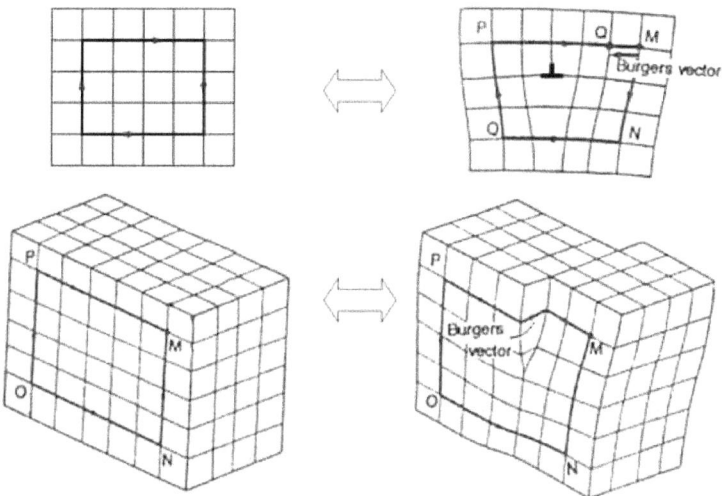

Figure 23. Schematics of edge (top) and screw dislocation (bottom) (image from wikipedia.org).

At the central point of a screw dislocation ($r = 0$), a singularity is formed whose theoretical stress is infinite, as

---

*56) The position has been shifted in the x direction by b for an edge dislocation and in the z direction for a screw dislocation.

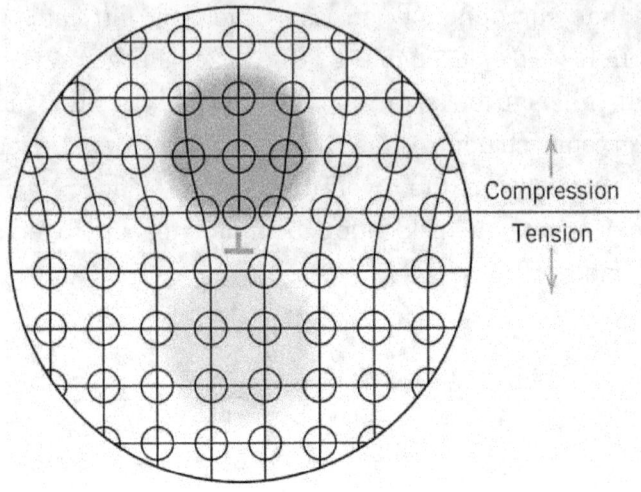

Figure 24. Stress distribution around an edge dislocation (image from academic.uprm.edu).

$$\tau_r = -\frac{Gb}{2\pi r} \quad \text{---} \quad (3.1),$$

where the shear stress τ_r is inversely proportional to the distance r from the center and is proportional to the shear modulus G and b. Practically, the stress at the central point is not infinite but has the highest level as shown in Figure 22. As shown in Fig. 24, the stress around an edge dislocation develops compressively on the upper part and in a tensile mode on the lower part around the mark (⊥) symbolizing the edge dislocation. This is because the arrangement of lattice below the horizontal line is one vertical line is missing compared to that above it.

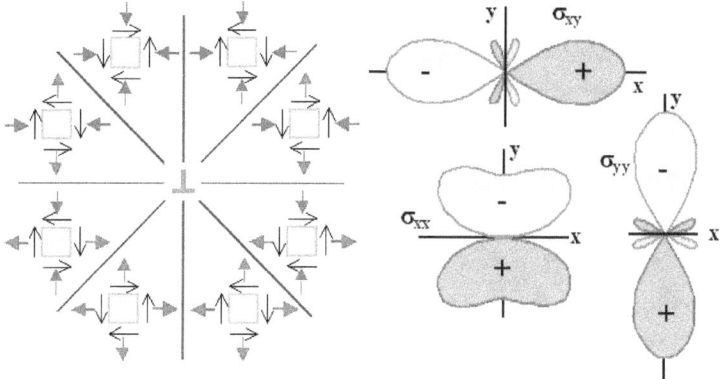

Figure 25. Type and distribution of stress around an edge dislocation (tf.uni-kiel.de)

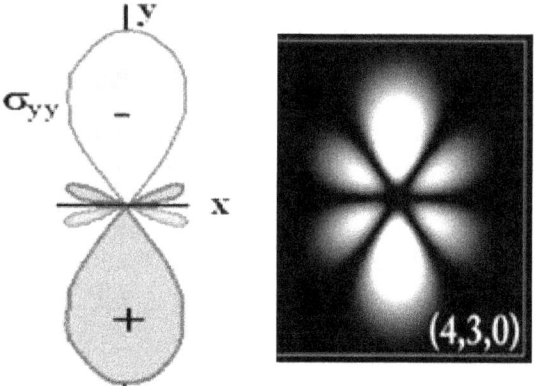

Figure 26. Stress field around an edge dislocation and an electron energy level of the hydrogen atom.

Electronic energy level

When viewed from the eight orientations around this edge dislocation, the stress field is more complex as shown in Figure 25.[62] These shapes are similar to the probability distribution of the electron of a hydrogen atom in Figure 16 (compared by selecting similar ones in these two figures in Figure 26). In this sense, the electrons in atoms are not really existing particles, but can be regarded as a kind of stress or strain fields generated when energy is stored in the vacuum lattice and the lattice point vibrates spherically or dislocates from the original point free of energy.

If the solid vacuum is a crystal, such as a metal with a regularly repeating lattice, the probability of finding an electron can be inferred from this stress field. The energy level of an individual electron of an atom has four indices: n (the principal quantum number), l (the azimuthal quantum number), m_l (the magnetic quantum number), and m_s (the spin quantum number). Two electrons can not have the same indices. n is a positive integer, being the quantum number representing the overall magnitude of the electron orbital. Figure 27 shows the energy levels corresponding to the principal quantum numbers for hydrogen. In the periodic table, the principal quantum number of valence electrons of an element is equal to its period number. Referring to Table 2, n represents the quantized distance of a plecton or PeN from the atomic center in the new atomic model, as PeNs stack in a sphere. The azimuthal quantum number ℓ (the angular or orbital quantum number) is a quantum number representing the angular momentum of electron in the atom and can have a value from 0 to $n-1$.

When the principal quantum number n is 3, it can have the value of ℓ = 0, 1, 2. It is chemically very important.[63] In the regime of the new atomic model, ℓ represents the number of the same kind of locations that a PeN can occupy when a new PeN is added on the surface of the atom. When a new PeN is stacked on the outmost shell of a completely spherical atom (e.g. Ar), it may be placed just above a PeN in the outmost shell (we call it a valence PeN instead of a valence electron) of Ar, or can be located between two valence PeNs, as the surface area increases with the increase in the radius (proportional to the square of the radius).*[57] Because these locations differ in energy, the total energy of the atom is different. The magnetic quantum number m_l is related to the number of locations for a PeN to occupy at the quantum number of ℓ. It can have a value from $-\ell$ to $+\ell$, including 0. Thus, the electron (namely PeN) configuration represented by s, p, d and f have the quantum numbers of 1, 3, 5 and 7, respectively. It is the number of locations of the same kind that a PeN can occupy for a given ℓ. Each quantum number can accommodates two PeNs. The electrons in these two PeNs have the opposite spin quantum number m_s. Electron spin is not physical one,[64] but is interpreted in terms of more abstract quantum mechanics*[58]

---

*[57] Referring to the two-dimensional atomic model in Figure 20, the number of locations a PeN can stack increases as the atomic radius (in this case, the radius of a circle) increases. For a two-dimensional atom, there are three types of locations: A, B, and C. This represents the azimuthal quantum number in the three-dimensional atomic sphere.

*[58] In the new paradigm, electron spin is simply and clearly understood if an electron is a circular standing electromagnetic wave.

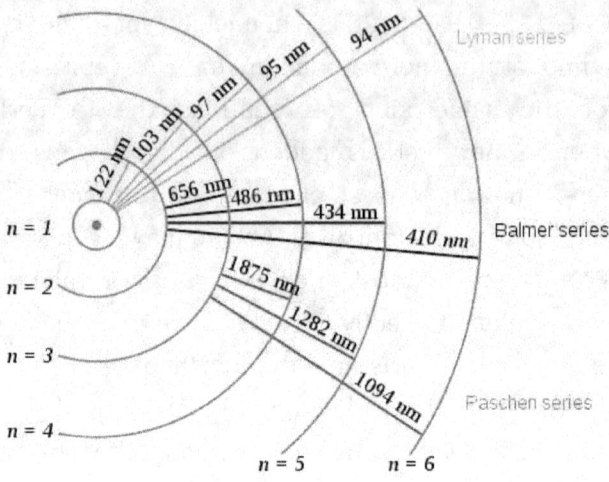

Figure 27. Locations of the electron in hydrogen corresponding to hydrogen absorption spectra (image from wikipedia.org). The location is determined by the frequency of the electron as a circular steading electromagnetic wave rather than indicating the specified location of the electron in the new vacuum paradigm.

Each element can be described by the four indices. Silicon (Si) with the atomic number of 14 has 14 electrons (i.e. 14 PeNs), so the electron configuration is $1s^2$ $2s^2$ $2p^6$ $3s^2$ and $3p^2$. Si is the addition of $3s^2$ $3p^2$ PeNs to the neon structure [Ne] (10, $1s^2$ $2s^2$ $2p^6$). s PeNs have no angular momentum, so the distribution is symmetric around the atomic center. p PeNs have non-zero angular momentum, so the distribution is asymmetric. So the addition of $3s^2$ PeNs of Si does not deteriorate the symmetrical stress field of Ne. It is assumed that the s^2 PeNs cause the existing atom to expand not occupying specific locations around the

Ne atom, or that the two PeNs roam the surface of the entire sphere. On the other hand, there are six positions denoted by *p* at which PeNs can be located (up/down, left/right and front/back). Thus, the $3p^2$ PeNs of Si will symmetrically be positioning at each other's opposition. If one part expands, the opposite part contracts, so it will form a stress field similar to the stress distribution around an edge dislocation shown in Figure 24. Therefore, the placement of all the PeNs ($3s^2 + 3p^2$) at the outmost shell of Si in a tetrahedral structure will mitigate the stress to some extent. Si has a crystal structure of regularly repeating tetrahedrons, like diamond, when present as a pure solid. This diamond structure minimizes stress in the surrounding solid vacuum and stable. All interatomic bonds are so formed as to minimize the stress of the surrounding solid vacuum, which is discussed in detail in the next section.

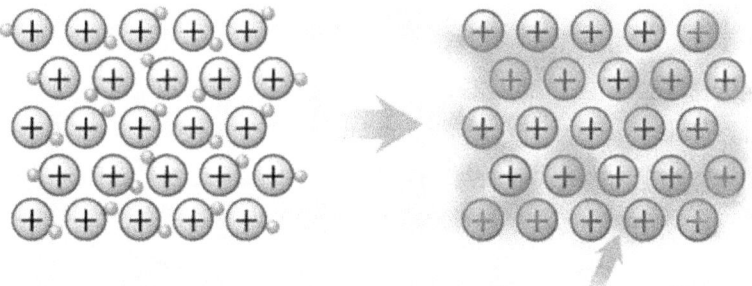

Figure 28. Metallic bonding model in modern physics. Valence electrons are not bound to specific ions, free to move and spread all over the metal ions (image from Assignment Point).

3.2. Interatomic bonding

According to current physics, when atoms combine to form molecules or crystals, there are three types of primary chemical bonds depending on the configuration of valence electrons of the atom: Ionic bonds, covalent bonds and metallic bonds.[65] Other type of bonds include hydrogen bonds and the van der Waals force. The latter are relatively weak and categorized as secondary bonds. In the atomic model of the new paradigm, not electrons but PeNs are involved in the bonding. Two PeNs combine to form a stable ^{4}He structure. In a covalent bond with hydrogen (except for hydrogen molecules), one neutron is deficient, so the combination of one PeN and one plecton give a ^{3}He structure. Although ^{3}He is a stable element, its ground state energy is higher than that of ^{4}He. So it is the interpretation of the new paradigm for the existence of the secondary bond called the hydrogen bond.*[59]

Metallic bonds

Metallic bonds in current physics are one type of chemical bonds caused by electrostatic forces between the free electrons with negative charges and the metal ions with positive charges. As shown in Figure 28, the valence electrons are dispersed rather than bound to specific metal ions. The cations share the electrons. The number of valence electrons in metallic bonds is insufficient for ionic or covalent bonding. It is a model where valence electrons

---

*[59] Please refer to Section 3.3 for details.

move almost freely between the metal ions. The physical properties of metals such as strength, ductility, electrical and thermal conductivity are macroscopic representations of the microscopic behaviors of these electrons and ions.66 In the new paradigm, the electrons belonging to an atom do not escape the protons in solids. The absence of free electrons in conductors is the primary premise of the new theory of electrical conductivity covered in this book. So what kind of bonds do metals? If valence electrons are replaced by valence PeNs in the new paradigm, it is assumed that a valence PeN in the s orbital is not localized at a specific position, but is positioning or moving with the same probability on the entire atomic surface, as mentioned in the previous section. If a PeN in the s orbital is located opposite to a PeN in the s orbital of the adjacent atom, the energy distorting the surrounding structure will be high, and vice versa. So we may think that the electrical energy in a conductor is stored in the form of (electromagnetic) strain energy, when an external electric field is applied, and an electric current is the propagation of the stored strain energy.*60)

First let's consider the element sodium (Na). Na has a valence electron of $3s^1$, only one PeN at its outermost shell. This PeN will be responsible for metallic bonding, as only one PeN and thus a plecton and a neutron participate in the bond.*61) In the new atomic model, the valence PeNs tend to form energetically stable ^{4}He units. In this context,

---

*60) Electrical conductivity in the new paradigm is covered in detail in Chapter 5.

*61) Increasing one plecton in the atomic periodic table inevitably increases one neutron.

chemical bonds can be understood in terms of the formation of ^{4}He structures. When one valence PeN of an atom and one PeN in the adjacent atom are shared, one ^{4}He unit is formed. If a PeN of the atom is shared with the eight adjacent atoms in total, there will be eight PeN units in the outermost shell, creating a perfect spherical atom. It can be understood that the crystal structure of all metallic elements in group 1A is bcc. If the number of the nearest adjacent atoms*[62] is 8 and they share all the valence PeNs from these atoms, the number of ^{4}He units in the outmost shell becomes eight. The shell is completed. However, since one PeN must be shared with the eight adjacent atoms, the bond strength will be low and the bond can be thought of as constantly moving. Namely the process of bonding and debonding will be dynamically equilibrated depending on temperature and pressure.

Let's look into magnesium (Mg) in the same element period. The valence electron configuration is $3s^2$. As two PeNs can participate in the bond, the modulus of elasticity of Mg is expected to be greater than Na (Mg: 45 GPa[67] vs. Na: 10 GPa[68]). The crystal structure is hcp. Normally, a high bonding force (between PeNs) is expected to reduce the conductivity because the bond is less flexible, but Mg has the electrical conductivity almost similar to or slightly higher than Na: for Na 2.1×10^7 S/m,[69] for Mg 2.3×10^7 S/m.[70] So, it can be seen that Mg becomes more dense in the crystal structure and the bond strength has increased due to the additional PeN.

Aluminum (Al) in the same element period has the

---

*62) It is called the coordination number.

configuration of $3s^2$ $3p^1$ and a fcc structure. More than three PeNs can participate in the bond, so the modulus of elasticity is greater (70 GPa)[71] than that of Mg with a hcp structure and the electrical conductivity is about 1.7 times that of Mg (3.8×10^7 S/m).[72] Some of the three PeNs contribute to primary bonds to form the crystallographic structure, while some contribute in the conductivity by forming flexible secondary bonds that easily react to external electric fields.*[63]

Ionic bonds

An ionic bond, as with NaCl, is a type of primary chemical bonds, mainly between a metallic atom (one PeN in Na in the $3s$ orbital) and a nonmetallic atom (7 PeNs in the outmost shell of Cl). In the regime of conventional physics, the valence electron of the metal atom ($3s^1$ electron of Na in NaCl) is donated to the outermost shell of the nonmetallic atom (the configuration of Cl becomes $2s^2$ $2p^6$ in NaCl). For NaCl, by donating the $3s^1$ electron of Na in NaCl to the outermost shell of Cl, the Na atom becomes a Na⁺ ion and the Cl atom becomes a Cl⁻ ion. NaCl can also be regarded as the combination of one Ne atom and one Ar atom in terms of the valence electron. The binding energy of this bond is mainly due to the electrostatic force formed between a pair of positive and negative ions, proportional to the magnitude of the two charges and inversely proportional to the square of the distance. NaCl has a binding energy of about 3.22 eV per atom.[73]

---

*63) This is discussed in detail in Chapters 5 and 6.

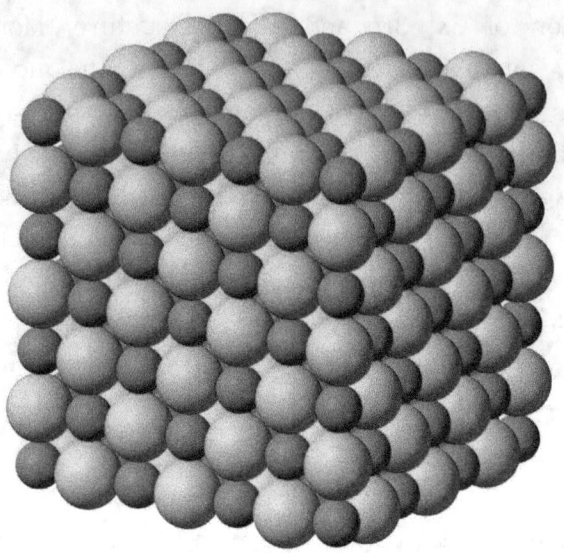

Figure 29. Crystal structure of NaCl (image from wikipedia.org)

In terms of the new atomic model, the $3s$ PeN of Na, combined with the 7 valence PeNs of Cl, appears to form the Ne structure [Ne] from Na and the Ar structure [Ar] from Cl, respectively. As the Na atom has only one valence PeN, the stress field in the solid vacuum by this PeN is strongly localized, while the Cl atom yields an unbalanced stress field as it lacks just one PeN in the completed 8-PeN shell structure. When the 凸 part of the Na atom and the 凹 part of the Cl atom meet, the stress imbalance is mitigated by forming a stable bond. Since even one NaCl molecule does not form a perfect sphere, the imbalance in the stress distribution in the surrounding solid vacuum is somewhat lowered by the aggregation of NaCl molecules to form a NaCl crystal. The NaCl crystal in Figure 29 does not distinguish between individual molecules called NaCl. It is thought that the interatomic bond does not belong to any particular atoms, but should be distributed evenly over a total of six bonds, so that individual ionic bonds are regarded as not fixed. This is similar to the case with metallic bonds, in which NaCl plays a role of a metal atom. Due to the nature of the *p*-orbital (of the Cl ion), ionic bonds exist in each of 6 locations in the direction of the *x*, *y*, *z* axis in a dynamic equilibrium,*[64]* and because NaCl has no extra PeNs, the crystal would not be very solid (the modulus is 40 GPa, greater than Na but less than Mg).[74]

NaCl in solid state is an insulator. This means that there are no flexible interatomic PeN bonds that can transmit energy in response to external electric fields. Ionic bonds themselves do not respond to external electric fields, because this bond is very rigid toward the external electric field, electrical energy cannot be stored or transferred.*[65]* In order for NaCl to form the structure shown in Figure 29, the valence 7 PeNs of Cl always accompany the valence PeN of Na, so that the 凸 part of the Na atom and the 凹 part of the Cl atom fits to make a stable bond. As this bond is very stable, it difficult to respond to external electric fields. On the other hand, electricity flows in solutions or in molten states. In conventional physics or chemistry, ionic bonds are separated (dissociated) into some anions and cations in aqueous solutions. A dissociated aqueous solution is characterized by its conductivity. In a NaCl solution, there is the same number of Na$^+$ ions (one

---

*[64] The process of bonding-debonding is in dynamic equilibrium.
*[65] Electrical conduction needs flexible chemical bonds because it is the transfer of the distortion energy mostly of secondary bonds. See Chapter 5.

electron lost) and Cl⁻ ions (one electron obtained). In the new atom model, Na⁺ is a state deprived of a PeN of Na. In this case the Na atom must become a neutral Ne atom. If this case does not occur, it can be thought that the ionic bonds are separated due to the presence of water molecules. In aqueous solutions, polar water molecules will intervene in the NaCl bonds, leading to an increase in the distance between the 凸 of Na and 凹 of Cl, thus increasing the flexibility of the bond.

As we will see in the next chapter, in the new paradigm, the electrical conduction is the movement of the excited state of secondary interatomic PeN bonds in conductors in the presence of external electric fields. In aqueous solutions, the ionic bond of NaCl have an increased flexibility via the intervention of water molecules. In a solid NaCl, there is no flow of electricity because of the inflexibility of this ionic bond so that it does not react to external electric fields. In contrast, ionic bonds that are pliable in an aqueous solution (or in a molten state) react enough quickly to external electric fields and transfer their bond energy in a wave form. This is why solids formed by ionic bonds increases its electrical conductivity in molten state or in aqueous solutions in the new paradigm.

Covalent bond

A covalent bond, also called a molecular bond, shares a pair of electrons between two different atoms donated by each atom, such as methane (CH_4) in Figure 30. Thus, each covalently bound atom appears to have eight valence

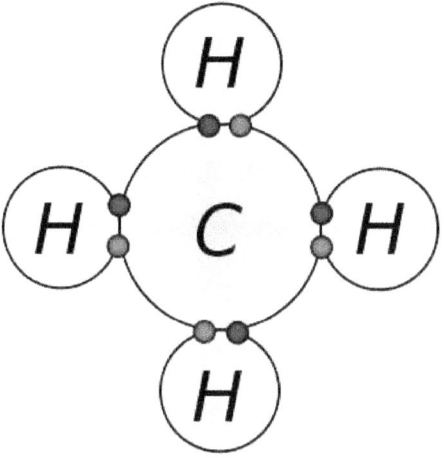

- ○ Electron from hydrogen
- ● Electron from carbon

Figure 30. In conventional physics, a covalent bond is a bond sharing one electron pair between the two adjacent atoms (image from wikipedia.org).

electrons (two for the hydrogen atom). The bonding force is generally larger than that of ionic bonds. Two hydrogen atoms also share electrons to form a stable hydrogen molecule.[75] Covalent bonds form well between the atoms with similar electronegativity. The Cl-Cl bond in Cl_2 is an example of the covalent bond between the same kind. Cl has the PeN structure of $3s^2\ 3p^5$, and seven ^{4}He units are already formed on the 8-PeN structure of Ne ($2s^2$, $2p^6$), and one more PeN is required to complete the 8-^{4}He structure. The Cl atom shares two PeNs with the other Cl atom to form a 2-PeN or ^{4}He structure, to have the [Ar] structure. According to the new paradigm, when the number of valence PeNs is eight, the stress field developed by the

8-PeN structure has a spherical symmetry and is thus stable. Covalent bonds are strong because they are highly localized. The conductivity is also very low because the bond is very stiff to respond to external electric fields to transfer electrical energy in a liquid or solid state.

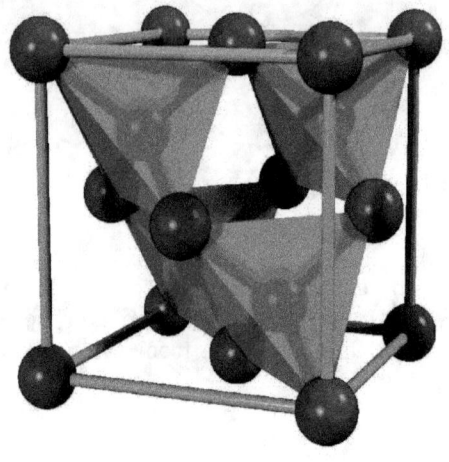

Figure 31. Diamond crystal structure (image from wikipedia.org)

Diamond is very representative in this regard. Diamond needs 4 more PeNs to have the 8-PeN structure in the outermost shell as with Ne. As shown in Fig. 31, the [Ne] structure is completed by sharing 4 PeNs with four adjacent carbon atoms. However, it is doubtful whether in the covalent bonds of diamond the 8 PeNs will have the [Ne] structure or 4 ^{4}He units are formed at the vertex of a tetrahedron. In the latter case, the very high modulus of elasticity of 1,220 GPa is understood, compared to its density (3.51 g/cm^3).[76] Since diamond does not have flexible bonds for electrical conduction, it is an insulator*[66] with the conductivity of 10^{-13} S/m (resistivity 10^{12} Ω·m @20°C).[77] On the other hand, graphite whose valence PeN configuration is not of [Ne], but has a two-dimensional structure as shown in Figure 32. One carbon atom uses 3 PeNs for covalent bonds with three adjacent atoms and the remained PeN is involved in electrical conduction. Because of its two-dimensional nature, the conductivity is anisotropic, 2~3×10^5 S/m in the basal plane and 3.3×10^2 S/m in the perpendicular direction to this plane. Compared to diamond, the conductivity is higher, while the elastic modulus is very low, 4.1 ~ 27.6 GPa.[78] On the other hand, Si of the same IVA group ([Ne] $3s^2\ 3p^2$) has the same crystal structure as diamond, but its resistivity is 2.3×10^3 Ω·m (@20°C),[79] which is significantly lower than that of diamond of 10^{12} Ω·m and the elastic modulus is 130 GPa to 188 GPa,[80] about one tenth of diamond. The density is 2.32 g/cm^3, which is lower than that of diamond. Si is an element in the period 3 that already has four 2-PeN (^{4}He) units (8 PeNs of the [Ne] structure + 4 valence PeNs results in 4 ^{4}He units and 4 unpaired PeNs on the surface of the [Ne] structure). Since these unpaired four PeNs are in the period 2 of the [Ne] structure, the length of the interatomic 2-PeN bond would be larger than that of diamond. Therefore, the covalent bond of Si is weaker than that of

---

*[66] Unlike most electrical insulators, diamond is a good thermal conductor. Strong covalent bonds and low phonon scattering are the reasons. The thermal conductivity is 2,200 W/(m·K), which is five times higher than silver, the metal with the highest thermal conductivity.

diamond, and it reacts more flexibly to external electric fields, so it can be understood that the elastic modulus or resistivity of Si is fairly different from those of diamond and the density is low.

Based on the new atomic and chemical bond models, even covalent bonds can lose their bond stiffness (momentarily) in response to an external electric field and thus the crystal structure based on covalent bonds may be distorted (interatomic PeN bonds are relatively displaced) to accommodate an electric current. It is similar to the case of a piezoelectric element, where electricity flows when the crystal structure is distorted.

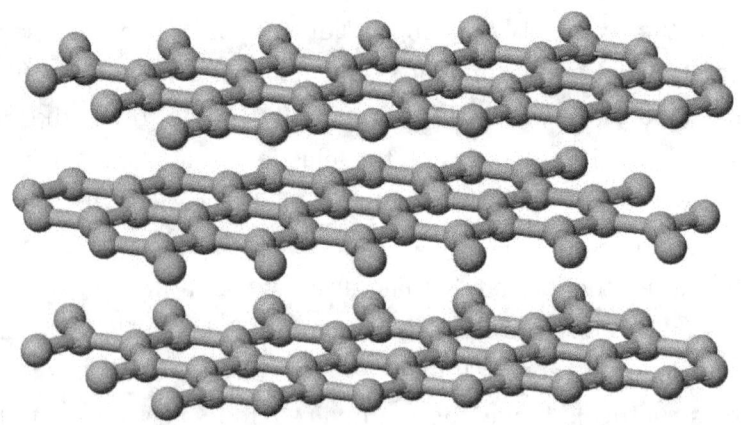

Figure 32. Graphite crystal structure (image from wikipedia.org)

For CH_4, the hydrogen atom provides one plecton, not a PeN, for a covalent bond. Thus the covalent bond in CH_4

does not give the ^{4}He structure, but the ^{3}He (a plecton + a PeN) structure, energetically less stable. Because of this, when an element other than hydrogen covalently binds with hydrogen, it is thought that different characteristics may appear because the hydrogen atom has no neutron. For this reason, a secondary bond called a hydrogen bond is formed in addition to the covalent bond of hydrogen compounds, which will be described in detail in the following section.

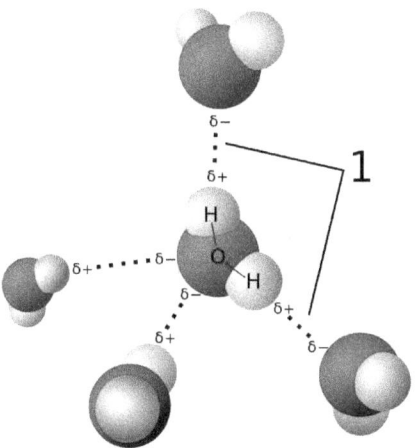

Figure 33. Hydrogen bonds between water molecules. Dotted lines indicate hydrogen bonds (image from wikipedia.org).

3.3. Hydrogen bond

A hydrogen bond is known to be a bond due to partial electrostatic forces and it is formed between a hydrogen atom covalently bound to an electrically more negative atom (oxygen, nitrogen, fluorine) and an adjacent atom other than hydrogen atom. A water molecule in Figure 33 is the most typical example. Hydrogen bonds are formed between molecules or within one molecule, and their energy ranges from 1 to 40 Kcal/mol, depending on the constituent atoms, molecular structure, and environment.[81] It is formed between inorganic molecules such as water and organic molecules such as DNA.*[67] The high boiling point of water (100°C) is due to the hydrogen bond. Hydrogen bonds play an important role in the polymer structure, such as in forming secondary and tertiary structures of proteins and nucleic acids by intramolecular hydrogen bonding.

Structure and energy of hydrogen bonds

Atoms with high electronegativity that are not covalently bound to the hydrogen atoms in hydrogen bonds are called proton acceptors, and those covalently bound to the hydrogen atoms are called proton donors. Referring to Figure 34, when hydrogen bonds are formed between water (H_2O) and ammonia (NH_3) molecules, two types of hydrogen bonds are formed: between the hydrogen atom in H_2O and the nitrogen atom in NH_3 (in this case the oxygen atom in H_2O is a proton donor) and between the hydrogen atom in NH_3 and the oxygen atom in H_2O (in this case the nitrogen atom in NH_3 is a proton donor).

---

*[67] Deoxyribo Nucleic Acid. As shown in Figure 39, two long strands of nucleotides wrap around each other to form a double helix structure.

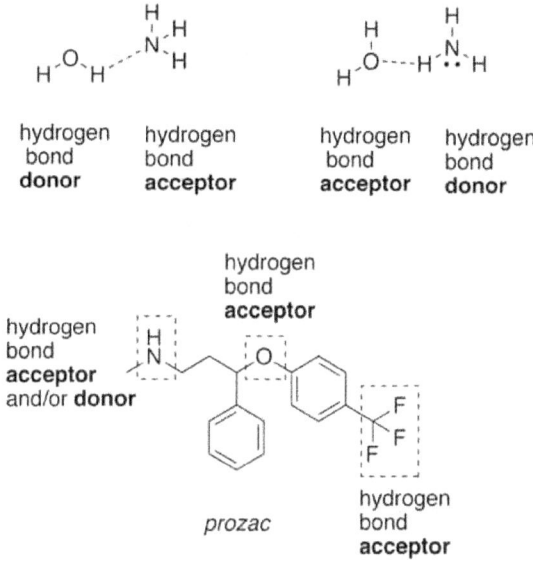

Figure 34. Acceptors and donors in hydrogen bonds (image from wikipedia.org)

A hydrogen bond is the interaction between two electric dipoles,[*68)] but it has characteristics of covalent bonding. That is, it is anisotropic and the bonding is quite strong. These characteristics become more pronounced when the proton acceptor has a higher electronegativity. The bond strength of a hydrogen bond varies from 1-2 kJ/mol to 161.5 kJ/mol.[82]

The length of covalent bond (X-H) in one molecule is usually 110 pm (pm = 10^{-12} m) and that of hydrogen bond

---

[*68)] It is the separation of positive and negative electric charges. An electric dipole is quantified by a vector called the dipole moment. Direction from a negative charge to a positive charge, expressed as the product of the charge strength and the distance between the charges.

(H⋯Y) between the two molecules is 160-200 pm. The hydrogen bond of water is 197 pm in length. Analyses with infrared spectrometers showed that the formation of a hydrogen bond results in a lower frequency of stretching vibration (and thus the weaker bond strength) or, conversely, the higher frequency and stronger bond strength that reduces the bond length.[83]

Table 3. Binding energy of various hydrogen bonds.

Chemicals	Hydrogen bond	bonding energy J/mol (kcal/mol)
HF_2^-, bifluoride	F-H⋯:F	161.5 (38.6)
water-ammonia	O-H⋯:N	29 (6.9)
water-water, alcohol-alcohol	O-H⋯:O	21 (5.0)
ammonia-ammonia	N-H⋯:N	13 (3.1)
water-amide	N-H⋯:O	8 (1.9)
	HO-H⋯:OH^{+3}	18 (4.3)

The hydrogen bond strength can be expressed by the non-covalent interaction (NCI) index. Unlike covalent bonds, this interaction does not share electrons,[84] and varies in the electromagnetic force in one molecule or between molecules. The NCI index can visualize the dispersed electron density and the density gradient as shown in Figure 35. Quantum chemistry uses this in various ways such as for the new drug development.[85]

Hydrogen bonds, not covalent bonds, are considered to be a type of the electrostatic force and the strength

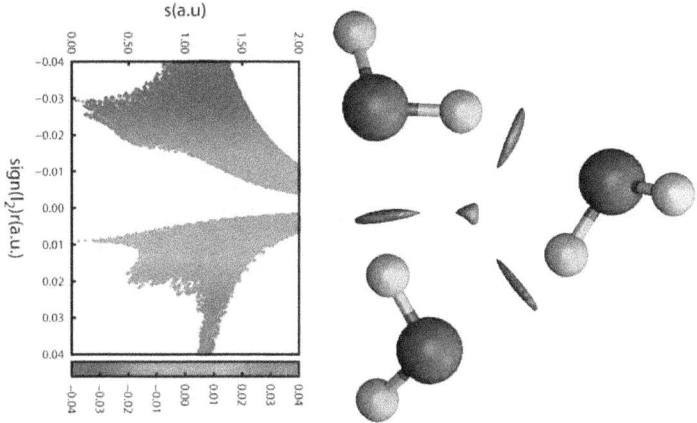

Figure 35. Visualization of the NCI Index in 2D and 3D for three water molecules (image from wikipedia.org).

depends on the intermolecular distance. Although there are differences in the energy from covalent and ionic bonds, hydrogen bonds are also a type of chemical bond. The net energy of a bond due to the interaction becomes negative, leading to a stable bond. Previously, hydrogen bonds were interpreted as having the characteristics of covalent bond, but there were many argues. It is stated that this explanation is possible only if NMR*[69] has the

---

*[69] Nuclear magnetic resonance is a resonance phenomenon in which an atomic nucleus placed in a strong magnetic field oscillates by a weakly vibrating magnetic field. It emits electromagnetic waves at a specific frequency of the nuclear magnetic field when the frequency of the external weak magnetic field matches the natural frequency of the nucleus. The natural frequency of the nucleus depends on the strength of the external magnetic field, the chemical environment, and the magnetism of the isotope. The frequency when the intensity

characteristics of covalent bond.[86]

Let's interpret the characteristics of hydrogen bond in terms of the atomic model of the new paradigm. As mentioned in the previous section, an additional secondary bond is necessary for a covalently bound molecule with hydrogen to be energetically more stable. A covalent bond with hydrogen results in the less stable ^{3}He structure than that of ^{4}He in the bond. This is the origin of hydrogen bonding in the new paradigm. Because molecules that covalently bind with hydrogen lack one neutron in their bond, a lower energy state can be obtained by sharing plectons of adjacent molecules that can fill the neutron-deficient sites. In this regard, hydrogen bonds can be said to have the characteristics of covalent bond. For an element with a high electronegativity, the asymmetric distortion formed in the surrounding solid vacuum is more severe than for an element with a low electronegativity. For this reason, it can be understood that the hydrogen bonding force is higher for an element with a higher electronegativity,

Hydrogen bonds in water

A representative example of molecules with hydrogen bonds are water molecules. As shown in Figure 33, the hydrogen bond is a intermolecular bond between the oxygen atom of

---

is about 20 tesla is practical because it corresponds to the TV broadcasting frequency (60-1000 MHz). NMR spectroscopy using this phenomenon is used to investigate the structure of organic molecules in solution and to study molecular physics, crystals, and amorphous materials.

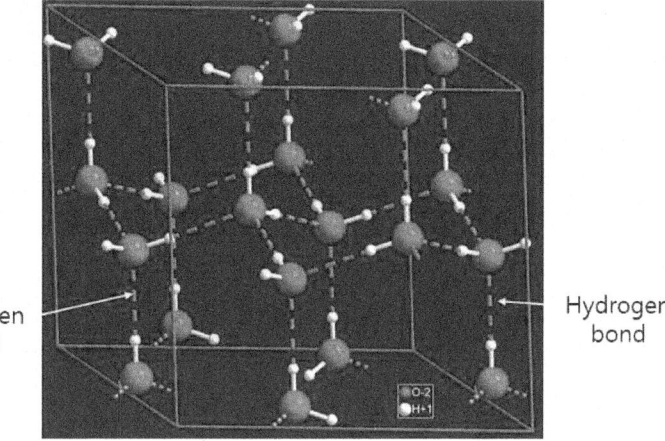

Figure 36. Hexagonal structure of ice. Dotted lines indicate hydrogen bonds (image from wikipedia.org)

one water molecule and the hydrogen atom of another water molecule. In a water molecule, two PeNs out of 6 valence PeNs of the oxygen atom are covalently bound to a pair of hydrogen atoms. Other two pairs of valence PeNs are capable of hydrogen bonding with the hydrogen atoms from other water molecules. Hydrogen fluoride has three pairs of valence PeNs, but one fluorine atom is capable of two hydrogen bonds.*[70]) In water molecules, hydrogen bonds repeat regularly three-dimensionally below the freezing point, forming ice crystals as shown in Figure 36. The high boiling point of water is owing to the high density of hydrogen bond and the bond strength. The viscosity is

---

*70) This is stated in "hydrogen bond" of wikipedia.org, but the bibliography is not provided so that we do not know the reason. As with water molecules, two hydrogen bonds are expected to make the molecule more symmetric and thus more stable against the surrounding solid vacuum.

also high compared to similar liquids. Oxygen in water molecules has two covalent and two hydrogen bonds with hydrogen. When water molecules form ice crystals, the maximum number of chemical bonds is four. The number of covalent and hydrogen bonds in liquid water depends on time and temperature.[87] The molecular motion becomes active and the density decreases with an increase in temperature, so the bond number decreases to 3.69 at 0°C, 3.59 at 25°C and 3.24 at 100°C. There is also a recent study suggesting that the combined bond number is 2.357 at 25°C.[88] The lifetime of hydrogen bonds between water molecules is around 10^{-11} seconds (10 picoseconds).[89] The hydrogen bonding structure in water can be studied by analyzing the vibration mode of O-H groups by infrared spectroscopy.[90]

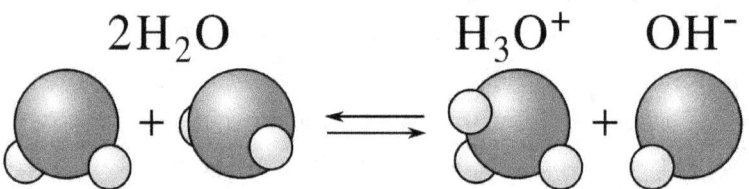

Figure 37. Autoionization or autodissociation of water (image from wikipedia.org). One hydrogen atom in the water molecule loses its nucleus and becomes a hydroxide ion OH⁻. The hydrogen nucleus H⁺ immediately combines with other water molecules to form a H_3O^+ ion. The dissociation constant is given by the product of the molar concentrations of each ion and is 10^{-14} at 25°C.

Sudden weakening of the hydrogen bonding force in the solid-state phase transformation in which the crystal structures change can be thought to be due to the change of orientation or to the rotation of the constituting ions.[91] In some compounds, the abnormal temperature range of phase transition from liquid to solid and back to liquid at higher temperatures is attributable to the characteristics of hydrogen bond.[92]

If the bond strength is constant, the hydrogen bond can be interpreted in terms of the two multi-atomic systems with opposite electric charges, as shown in Figure 37, instead of the constituent atoms of two interacting water molecules. In other words, OH^- (hydroxide) is an anion and H_3O^+ (hydronium) is a cation, which is interpreted as a bond between two ions having the opposite electric charges.[*71] For pure water at the standard condition (25°C, 1 atmosphere) this approach of cation and anion is rarely applied. On average, however, water molecules provide protons (hydrogen nuclei) to other water molecules in the ratio of 5.5×10^8. This is consistent with the dissociation (ionization) constant of water.

When water plays a role of the solvent in a solution, it forms a hydrogen bond between the proton donor of the solute and the acceptor of the solvent, but completely blocks the formation of hydrogen bonds between or in solute molecules. Consequently, hydrogen bonding between

---

*71) The new atomic model assumes that there is no separation of a plecton into an electron and a proton in solids or liquids. An electron-lost hydrogen atom is a proton and its size is negligibly small compared to the hydrogen atom. In this case, the distortion of the solid vacuum will also be minimal. It is unimaginable to engage in any bonding.

solute molecules is less preferred than that between water molecules and solute molecules.[93]

Hydrogen bonds have a profound effect on the three-dimensional structure of polymers. When hydrogen bonds occur between certain parts of a macromolecule, as shown in Figure 38, the molecule forms a specific structure, which plays a very important role in molecular physiology and biochemistry. For example, the double helix structure of DNA in Figure 39 is usually due to hydrogen bonding between a pair of underlying skeletal molecules. Hydrogen bonds play an important role in the stabilization of the tissues of protein molecules.[94] Wool is a protein fiber that is stretched and then pulled away when there is no external force due to the presence of hydrogen bonds. Washing wool at high temperatures permanently loses hydrogen bonds.

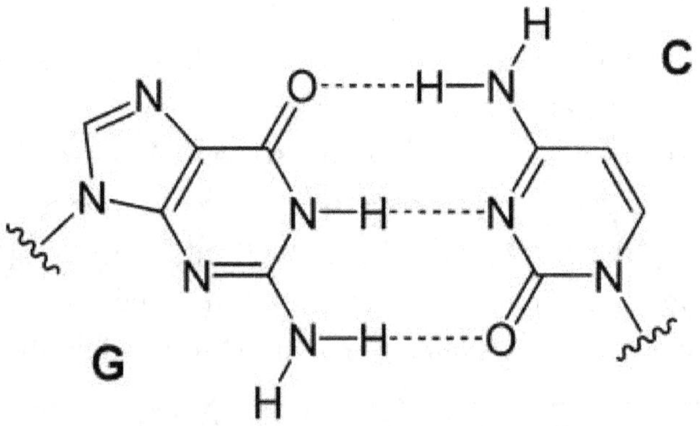

Figure 38. Hydrogen bonds between guanine (G) and cytosine (C), It is one of the two types of the base pairs in DNA (image from wikipedia.org).

Hydrogen bonds in polymers

One hydrogen atom can be involved in two hydrogen bonds. This is called the bifurcated state, which exists in organic molecules. Bifurcated hydrogen atoms are the key to the rearrangement of water molecules.[95] Acceptor-type hydrogen bonds (which terminates at the electron pair of oxygen) tend to easily form bifurcation compared to donor-type hydrogen bonds. This is called OCO (overcoordinated oxygen).[96]

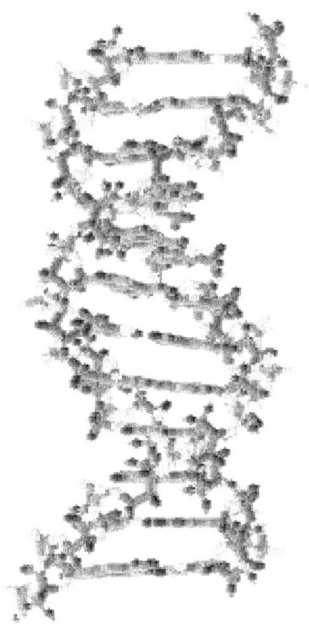

Figure 39. DNA double helix structure (image from wikipedia.org)

Symmetric hydrogen bonds are a special form of hydrogen bonds in which the proton (hydrogen nuclei) is exactly centered between the two identical atoms.*72) The hydrogen atoms are partially covalently bound to the two atoms. Therefore, the bonding force is much stronger than ordinary hydrogen bonding. The bond order is 0.5, and it is comparable to that of an ordinary covalent bond. It is also observed in highly pressurized ices, solid anhydrides (fluoric acids, formic acids, etc.) and liquid fluoride ions [F-H-F]⁻. In a formic acid, symmetric hydrogen bonds were observed at high pressures above 10,000 atm.[97] If the interatomic distance is very short, the activation energy of the hydrogen bond is small. This is called the low-barrier hydrogen bond (LBHB).[98] The length of a standard hydrogen bond is long (2.8 Å for an O...O hydrogen bond) and the hydrogen ions belong to heteroatoms. Very similar acid dissociation constants pK_a*73) of heteroatoms allow LBHBs at short distances (~2.55 Å). If the distance is further reduced (<2.29 Å), the hydrogen bond becomes a single-well or short-strong hydrogen bond.[99]

A hydrogen bond is a type of bond that is additionally formed between covalently bound molecules with hydrogen. This additional bond occurs because a hydrogen-based

---

*72) In the new atomic model, a proton is substituted by a plecton for this case.

*73) The acid dissociation constant, denoted by K_a, is a measure of acidity in a solution. This is the equilibrium constant of a chemical reaction known as the dissociation in acid-alkali reactions. When a compound HA dissociates into A⁻ and H⁺ ions in a solution and is in equilibrium, K_a is given by

$$K_a = \frac{[A^-][H^+]}{[HA]}$$

It is convenient to use its log value pK_a.

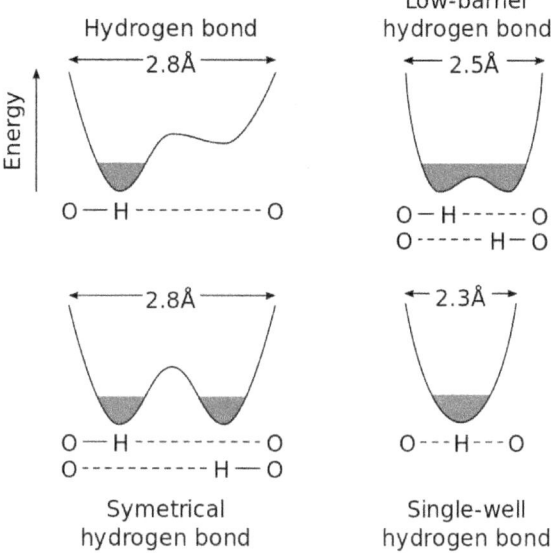

Figure 40. Energy distribution according to four hydrogen bonding states (image from wikipedia.org).

covalent bond yields the ^{3}He bond structure, being less stable than that of ^{4}He in other normal covalent bonds. In addition, elements that participate in hydrogen bonding are in the second period such as nitrogen, oxygen and fluorine. Referring to Table 2, in the new atomic model, the elements in the second period form the base layer of 8-PeNs for the double-layered shell[*74] at $r = 2$. In this sense, the elements in the second period could be more active than those in the third period, which can complete

---

[*74] The first layer of the double-layered shell composed only of ^{4}He is completed into [Ne] and the second layer is completed into [Ar].

the 8-^{4}He shell structure [Ar] with additional 8 PeNs. If the oxygen atom with 6 valence PeNs in a water molecule receives additional 2 PeNs, its outermost shell will be filled up, and be the base layer with 8 PeNs. But if one oxygen atom forms two ^{3}He covalent bonds with two hydrogen atoms, the remaining two pairs of valence PeNs (4 PeNs) can make 2 ^{4}He units by themselves. Hence the outer layer is composed of 2 ^{3}He + 2 ^{4}He units. In this case, two valence PeNs have to move to the locations of the third period, which requires energy that much. The energy can be supplied by hydrogen bonding with other water molecules with ^{3}He covalent bonds. The result will be a transformed ^{4}He bond structure, not the perfect ^{4}He one.

As stated above, the fluorine atom in HF has seven valence PeNs in the base layer, but the atom is only capable of two hydrogen bonds. The remaining three pairs of PeNs in the atom can constitute three ^{4}He structures. Three pairs of PeNs do not participate in hydrogen bonding, and only one PeN of the fluorine atom in HF makes two hydrogen bonds to make a distorted ^{4}He unit. Hence, the 4 ^{4}He bond structure is completed with this distorted ^{4}He unit for HF by hydrogen bonding. This is the picture of hydrogen bonding in the new atomic model. In other words, hydrogen bonding is inevitable that appears in the outermost layer making the bond composed of four ^{4}He units. It is thus understood that the electrical conductivity of ice is very low because four covalent and hydrogen bonds are tightly cross-linked like diamond.

3.4. Van der Waals Force

The van der Waals[*75)] force is a kind of inter-atomic or -molecular interaction. It is an attractive force, but also includes a repulsive one between atoms, molecules, and surfaces. When the interatomic distance decreases below the critical point, the repulsive force begins to increase and the van der Waals force becomes repulsive. This distance is called the van der Waals contact distance and is known to be due to the repulsive force between the electron clouds in the atoms.[100] It is mainly caused by the vibration of particles in close proximity, which is different substantially from covalent or ionic bonds.[101] The force is very weak, between 0.4 and 4 kJ/mol, but when this force is summed up it can alter the structure. This is one of the fundamental forces in macromolecular chemistry, structural biology, polymers and solid-state physics.

Origin of the van der Waals force

Unlike ionic or covalent bonds, the van der Waals forces are not originated from chemical bonds in which electrons are directly involved. They are relatively weak forces and sensitive to external conditions. These forces are mainly the result of the temporal change in electron density. When the electron density is temporarily biased to one side of the nucleus, nearby atoms are attracted or rejected. At an

---

[*75)] Johannes Diderik van der Waals (11.1837 - 3.1923) was a Dutch theoretical physicist. He developed an equation of state for gas and liquid, for which he was awarded the Nobel Prize in Physics in 1910.

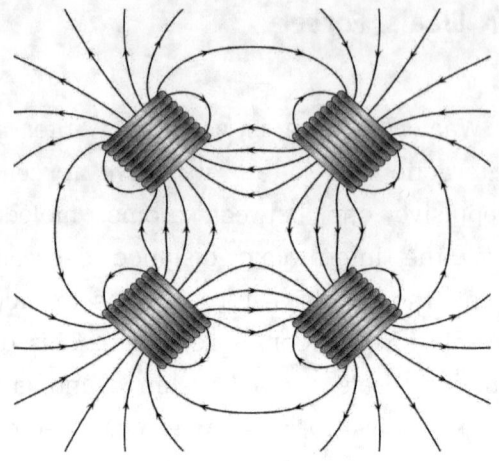

Figure 41. Magnetic field induced by quadrupole coil (image from wikipedia.org).

interatomic distance of 0.6 nm, the forces become very weak and below 0.4 nm they are repulsive. The four factors that govern the van der Waals forces are:

- Repulsion due to Pauli's exclusive principle to prevent the collapse of molecules.
- Attraction or repulsion between permanent charges, dipoles, quadrupoles (see Figure 41), and generally between permanent multipoles. This electrostatic interaction is also called the Keesom[*76)] force.
- Attraction between a permanent multipole of one molecule and induced multipoles in other molecules. This is

---

[*76)] Willem Hendrik Keesom (6.1876 - 3.1956) was a Dutch physicist. In 1926, he invented a method of making liquid helium. In 1921, he developed first a mathematical model of the interaction between electric dipoles. This is called the Keesom interaction after his name.

also called the Debye*77) force.
- Attraction between all pairs of molecules resulting from transient multipole interactions. Commonly referred to as the London*78) dispersion interaction.
-

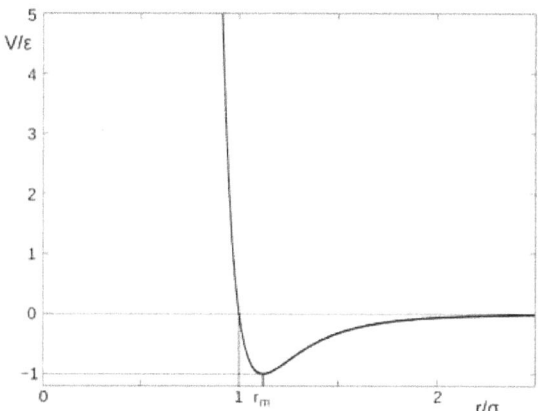

Figure 42. Graph showing a Lennard-Jones potential (image from wikipedia.org).

The van der Waals forces acting between all the molecules depend on the relative orientation of the

---

*77) Peter Joseph William Debye (3.1884. - 11.1966) was a Dutch-American physicist and chemist. In 1936, he won the Nobel Prize in Chemistry. In 1912, he developed an equation relating temperature and dielectric constants to the charge distribution in asymmetric molecules. In 1923, he developed a theory for the Compton effect that the frequency changes when x-rays interact with electrons.

*78) Fritz Wolfgang London (3.1900 - 3.1954) was a German physicist. He contributed greatly to the theory of chemical bonds and intermolecular forces. He developed with his brother Heinz the London equation to understand superconductivity. He was nominated for the Nobel Prize but did not win.

molecules. That is, they are anisotropic (except for inert gases). The Debye and London interactions are always attractive in any directions. However, the electrostatic force is attractive or repulsive depending on the molecular orientation because the charge sign changes when the molecule rotates. When a molecule thermally vibrates in a gas or liquid state, the electrostatic force loses its anisotropy. Thus the Keesom force is temperature sensitive. The effect of thermal vibration on the attractive force by induction or dispersion is relatively low.

Theory of the van der Waals force

The van der Waals force can be described for isotropic particles as a function of the distance between neutral atoms or molecules by the Lennard-Jones[*79)] potential (the interatomic potential), as[102]

$$V_{LJ} = 4\epsilon \left[\left(\frac{\sigma}{r}\right)^{12} + \left(\frac{\sigma}{r}\right)^{6} \right] = \epsilon \left[\left(\frac{r_m}{r}\right)^{12} - 2\left(\frac{r_m}{r}\right)^{6} \right] \quad \text{---} \quad (3.2),$$

where ε is the minimum energy, σ is the finite distance at which the potential energy is zero, r is the distance between the two particles. r_m is the distance at which the potential energy is the lowest. At r_m, the potential energy is $-\varepsilon$. $r_m = 2^{1/6}\sigma \approx 1.122\sigma$ and can be determined

---

[*79)] Sir John Edward Lennard-Jones (10.1894 - 11.1954) was a British mathematician and professor of theoretical physics at Bristol and Cambridge University. He is regarded as the founder of modern computational chemistry. He also proposed the Lennard-Jones potential in 1924.

experimentally or by quantum chemical calculations.

Among the van der Waals forces, the Keesom interaction, i.e. the electrostatic interaction between molecular ions, dipoles, quadrupoles and permanent multipoles, is the most prominent.[103] This force is an attractive one between the permanent dipoles sensitive to temperature changes.[104] The Keesom interaction occurs between molecules with permanent dipole moments, that is, between two polar molecules. The interaction does not form in electrolytic liquids.

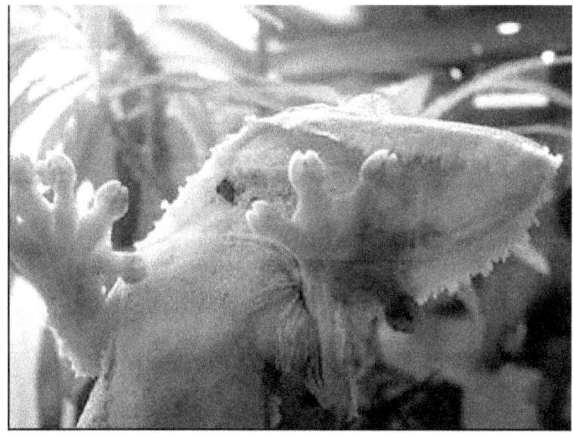

Figure 43. It was once known that the lizard is attached to the glass mainly due to the van der Waals force (image from wikipedia.org).

It is assumed that molecules do not settle and rotate constantly. But it can be settled by the Keesom interaction. The interaction energy of two fixed dipoles is inversely proportional to the third power of the distance, while the interaction energy is inversely proportional to the six power of the distance. On average, the energy V is given as

$$V = \frac{-m_1^2 m_2^2}{24\pi^2 \epsilon_0^2 \epsilon_r^2 k_B T r^6} \quad \text{---} \quad (3.3),$$

where m is the dipole moment, ε_0 is the vacuum permittivity, ε_r is the dielectric constant*[80] of matter, T is the temperature, k_B is the Boltzmann constant,*[81] and r is the intermolecular distance.

The second contributor to the van der Waals forces is the Debye force. This force is due to the interaction between a rotating permanent dipole and the induced dipoles, which are induced when one molecule with a dipole repels the electrons of other molecules.[105] Molecules with dipoles induce dipoles in the surrounding molecules, creating intermolecular attractive forces. The Debye force is not an interatomic force and is not temperature-dependent.

---

*[80] In electromagnetism, the dielectric constant or permittivity ε is an electrostatic capacity that a medium can have when there is an external electric field. In other words, it is a measure to store the electric field in the medium. It is the amount of charge for a material to produce the unit electric current, F/m. The vacuum permittivity is $\varepsilon_0 \sim 8.85 \times 10^{-12}$, and the permittivity of matter is often expressed as the ratio $\varepsilon/\varepsilon_0$, the relative permittivity.

*[81] It is a physical constant that correlates the temperature of a gas with the average kinetic energy of particles in the gas. $k_B = R/N_A$, where R is the gas constant and N_A is the Avogadro constant. Planck introduced it but borrowed the name Ludwig Boltzmann (2.1844. - 9.1906). He was an Austrian physicist who made a significant contribution to the development of statistical mechanics. The properties of mass, charge, structure, etc. of atoms are statistically analyzed to explain and predict the macroscopic properties of matter (viscosity, thermal conductivity, diffusion, etc.).

Induced dipoles can move freely and rotate around molecules without polarization. The Debye induction and the Keesom anisotropic effects are called the polar interactions.

An example of these forces is the interaction between hydrochloric acid (HCl) and argon (Ar). Induced dipoles are formed when the electrons in Ar are attracted to H or repelled by Cl in HCl. This interaction is expressed as

$$V = \frac{-m_1^2 \alpha_2}{16\pi^2 \epsilon_0^2 \epsilon_r^2 r^6} \quad \text{---} \quad (3.4),$$

where α is the polarizability.[*82] This interaction is smaller than the interaction between permanent dipoles, but stronger than the London dispersion.

The London dispersion is a weak intermolecular force, which occurs between temporal multipoles in the molecules of no permanent multipoles. This force is also called the transient dipole-induced dipole force. The strength of the interaction is proportional to the polarizability of molecule, which in turn depends on the total number of electrons and the area in which they are spread. Hydrocarbons have a small influence on this force, but this force increases when elements of high polarizability are present.[106]

[*82] The polarizability is a relative distribution of charges in materials in response to an external electric field. In a uniform medium α is the ratio of the induced dipole moment to the external electric field, i.e. $\alpha = p/E$. The SI unit is $C \cdot m^2 \cdot V^{-1}$ = $A^2 \cdot s^4 \cdot kg^{-1}$ and the cms unit is cm^3. In general, the polarizability of matter is anisotropic, and the polarizability of this anisotropic medium is represented by a tensor or a 3×3 matrix. A higher value of α_{yx} means that the polarization occurred strongly in the y direction in response to the electric field applied in the x direction.

For objects with the known number and volume of the constituting atoms and molecules, the total amount of the van der Waals forces is calculated by adding the forces of each interacting pair. Therefore, this force also depends on the shape of particles. For example, the van der Waals interaction energy between two spherical objects with radius R_1 and R_2 was calculated by Hamaker[*83)] in 1937 as:[107]

$$U=-\frac{A}{6}\left[\frac{2R_1R_2}{z^2-(R_1+R_2)^2}+\frac{2R_1R_2}{z^2-(R_1-R_2)^2}+\ln\left(\frac{z^2-(R_1+R_2)^2}{z^2-(R_1-R_2)^2}\right)\right]$$
--- (3.5).

A is the constant for materials characteristics, being negative or positive. It is called the Hamaker coefficient and has a value of 10^{-19}~10^{-20}. $z = R_1+R_2+r$ (r = the minimum distance between the two surfaces). In the case of $r \ll R_1$ or R_2, Eq. (3.5) is simplified as

$$U=-\frac{AR_1R_2}{6(R_1+R_2)r}$$ --- (3.6).

In other words, the van der Waals force is proportional to the harmonic mean[*84)] of the radius of the two objects and are inversely proportional to the distance. In this case, the force applied to the object is negative and obtained by

---

*83) Hugo Christiaan Hamaker (3.1905 - 9.1993) was a Dutch scientist. He presented a theory explaining the van der Waals forces between objects larger than molecules.

*84) The harmonic mean is the inverse of the arithmetic mean of the inverses of the given values. It is mainly used to find the average rate of change.

differentiating the potential energy, that is,

$$F_{VW}(r) = -\frac{dU(r)}{dr} = -\frac{AR_1R_2}{6(R_1+R_2)r^2} \quad \text{--- (3.7)}.$$

Van der Waals forces are evident among very fine dry powders of no capillary actions. These forces cause the powders to agglomerate and reduce fluidity. The adhesion by the van der Waals forces depends on the surface geometry. If the surface is rough or ridged, the surface area becomes larger, which increases the strength of van der Waals force. The above theory was calculated for individual pairs only and ignored the interactions and the delay effects of multiple particles. For a theory that reflects this, see the references.[108]

As with hydrogen bonds, the van der Waals forces are forces (or a kind of bonds) that act in molecules or solids. As metallic, ionic, or covalent bonds are primary chemical bonds, these forces are type of secondary bonds. In terms of the new atomic model, hydrogen bonds can be understood as additional bonds to compensate for the energy differences arising from the hydrogen-based covalent bonds. The van der Waals forces are also understood as forces in which two or more atoms are involved. Even if an element is made of all ^{4}He units, such as argon (Ar, 9 ^{4}He units), an inert gas, the surface is not perfectly spherical. Thus, since the distortion of the surrounding solid vacuum is not completely (point) symmetric but locally asymmetric, Ar becomes a solid at cryogenic temperatures (below −189.34 °C or 83.81 K). Molecules made of heteroatoms suffer more severely from such asymmetrical distortions of the solid vacuum. In this context, the van der Waals forces should be bigger for complex molecules. One example is the Keesom interaction, which is the electrostatic force generated between dissimilar elements. These force usually increases with a decrease in temperature as shown in Eq. (3.3) and will have a significant effect on the mechanical and electromagnetic properties of solids at low temperatures.

Regardless of the type of chemical bonds, energy is required to break or modify the bond structure. The types of energy supplied from the outside are mechanical (tension, compression, shear), thermal, or electrical ones. When mechanical forces are applied, primary bonds are mainly deformed or disintegrated, and the input of electrical energy will mainly result in the deformation or deterioration of secondary bonds. Thermal energy may be involved in both primary and secondary bonds. When electrical energy is applied by an external electric field, weak bonds (e.g. bonds due to electric dipoles interactions, the Keesom interactions) in a solid will be broken or distorted, and these deformed states propagates in conductors. In Chapters 4 and 5, the electromagnetic properties of solids are discussed and newly interpreted in terms of the deformation and propagation of secondary bonds, including the van der Waals forces.

IV. Electromagnetic property of solids

There are four fundamental forces in nature: the strong force, weak force, electromagnetic force and gravitational force. The strong and weak forces are only used within atomic nuclei. In solid-state physics, the electromagnetic force is the dominant one, and almost all the macroscopic properties of solids are interpreted in terms of electromagnetism. Classical physics deals with electricity and magnetism separately, but modern quantum mechanics integrates them into electromagnetism. Magnetism is always generated where electricity flows, and electricity is created through the changes in magnetism. The question, "what is electricity?", should be the same as the question, "what is magnetism?". This chapter summarizes in the regime of current physics what is the interaction between an electric field and matter, a magnetic field and matter. More precisely, it is summarized how the motion of electron and its accompanying magnetic field affect matter, especially solids, and how the electron motion and magnetism change in response.

4.1. Electromagnetism

As a result of Einstein's theory of special relativity, electricity and magnetism are fundamentally connected. Magnetism without electricity and electricity without magnetism are impossible. Electromagnetic theory combining electricity and magnetism is said to be perfectly consistent with the theory of special relativity.[109] In the theory of special relativity, electricity and magnetism are handled as a phenomenon and inseparable from each other. They are called electromagnetism.

According to the theory of relativity, the division of the electromagnetic force into the electric force and the magnetic force is not absolute but varies according to the coordinate system that depends on the observer. An electric force felt by one observer can be felt by another observer (another coordinate system) as a magnetic one or a combination of the two forces. Formally, in the theory of special relativity, the electric and magnetic fields are called electromagnetic tensors. If the coordinate system changes, these two components will be mixed.

Magnetic field (B, H and M)

A particle with charge q in an electric field **E** feel a force **F** corresponding to q**E**. However, if the particle is near an electrically conducting conductor, it also depends on its velocity v under the influence of the magnetic field. This force is called the Lorentz force and is expressed as follows.

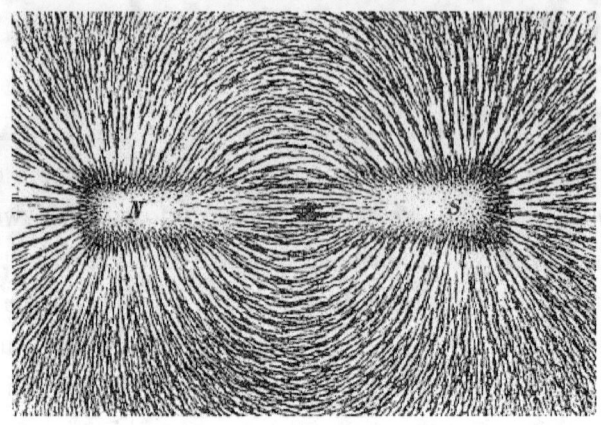

Figure 44. Iron particles arranged along the magnetic lines of a bar magnet (image from wikipedia.org).

$$\vec{F} = q(\vec{E} + \vec{v} \times \vec{B}) \quad \text{---} \quad (4.1),$$

where × means the vector product. B(=$\vec{B}$) is an (external) magnetic field and a vector. The introduction of vector allows accurately to describe the motion of charged particles. On the other hand, when matter is involved, not the vacuum, the magnetic field is denoted by H(=$\vec{H}$). In the vacuum, B and H are proportional to each other (substantially the same). They differ from each other in non-vacuum media. "Magnetic field" historically means H, and B was called in other terms. In recent physics textbooks the term 'magnetic field' sometimes means B and is used instead of H. For some media, magnetic field B is also known as magnetic flux density, and H is also called magnetic field strength. There is a relationship between the two:

$$\vec{B} = \mu_0 (\vec{H} + \vec{M}) \quad \text{---} \quad (4.2),$$

where the proportionality constant μ_0 is called the vacuum permeability. The magnetization $\mathbf{M}(=\vec{M})$ of the solid vacuum is zero. If the medium is not the solid vacuum, $\mathbf{M}$ is not zero and $\mu_0\mathbf{M}$ is called the magnetic polarization. The relationship between $\mathbf{B}$, $\mathbf{H}$ and $\mathbf{M}$ is shown in Figure 45. For small $\mathbf{H}$, $\mathbf{M}$ is proportional to $\mathbf{H}$ in diamagnetic and paramagnetic materials. That is, $\mathbf{M} = \chi\mathbf{H}$. χ is the proportionality constant called the magnetic susceptibility. In ferromagnetic materials, $\mathbf{M}$ is not proportional to $\mathbf{H}$ and nonzero even when $\mathbf{H}$ is zero.

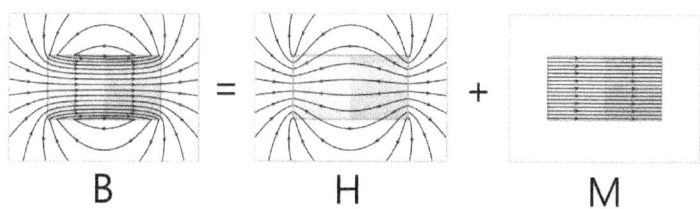

Figure 45. Relation between $\mathbf{B}$, $\mathbf{H}$ and $\mathbf{M}$ (image from wikipedia.org).

$\mathbf{M}$ is a vector quantity and represents the extent to which a material magnetizes in a zone to an external magnetic field, defined as the net magnetic dipole moment per unit volume. The strength of magnetization of a homogeneous magnet is a material constant, the magnetic moment divided by the volume. The SI unit of the magnetic moment is $A \cdot m^2$ and the SI unit of $\mathbf{M}$ is A/m, which is the same as the unit of $\mathbf{H}$. The (induction) magnetic field $\mathbf{M}$ in a zone is similar to the Ampére*85) law and has the following relationship to the current I_b flowing in a closed loop.110

$$\oint \overrightarrow{M} \cdot d\ell = I_b \quad \text{--- (4.3)},$$

where ℓ is the length of the closed loop.

Electricity has positive or negative charges, but magnetism has no such (magnetic) charges. However, in the isolated magnetic pole model, when one zone has a purely positive magnetic charge strength (for a N pole), it can be interpreted that the magnetic field lines goes into it rather than coming out from it. In this case, Eq. (4.3) can be modified to

$$\oint_S \mu_0 \overrightarrow{M} \cdot d\overrightarrow{A} = -q_M \quad \text{--- (4.4)},$$

where the integration is for the closed surface S and q_M is the magnetic charge in S. This is because the negative sign is directed from the S pole to the N pole. However, no independent magnetic monopole has yet been found, so Eq. (4.4) is only a mathematical expression for calculation.

The SI unit of **B** is tesla,*86) T (V·s/m^2) and of the magnetic flux Φ_B is weber,*87) Wb. Therefore, the magnetic

---

*85) André-Marie Ampére (1.1775 - 6.1836) was a French physicist, mathematician. One of the founders of classical electromagnetism. He invented solenoids and telegraphs.

*86) Nikola Tesla (7.1856 - 1.1943) was a Serbian American inventor, electrical and mechanical engineer. Known as the designer of modern alternating current electricity supply systems.

*87) Wilhelm Eduard Weber (10.1804 - 6.1891) was the first

flux density 1 Wb/m² is 1 T. Tesla's 10,000th is gauss.[*88)]
G. **H** is A/m in SI units and oersted Oe in CGS units.

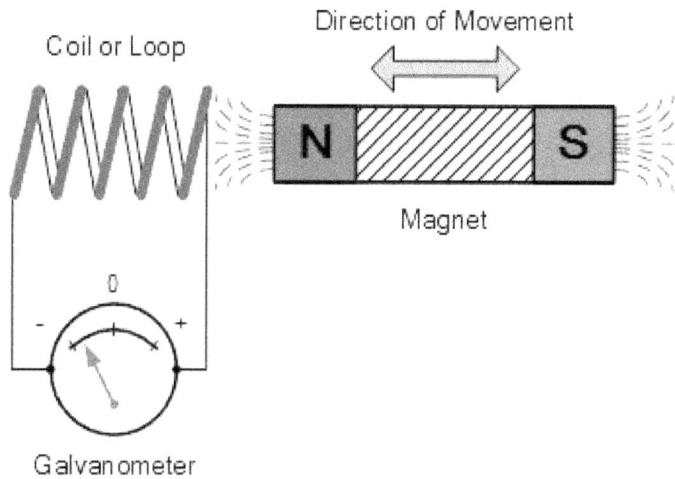

Figure 46. When a magnet moves inside the conductor coil, an electromotive force is generated according to Faraday's law (image from wikipedia.org).

Relation of electricity and magnetism

As a magnet moves near or inside a conductor coil, as shown in Figure 46, the magnetic field changes and in accordance with it an electric field forms and electricity

---

German physicist to invent electromagnetic telegraphs with Gauss.

*88) Johann Carl Friedrich Gauss (4.1777 - 2.1855) was a German mathematician, physicist. His outstanding contributions in various fields of mathematics and physics have earned him the title of one of the best mathematician.

flows through the coil. This phenomenon is called the electromagnetic induction and is summarized in the Lenz law. A current is induced in the direction to oppose the change in the magnetic field around the coil. Faraday made out of this phenomenon a law called Faraday's law of induction: the electromotive force (EMF) is proportional to the rate of change of Φ_B (magnetic flux) across the coil surface and the coil's winding number n. It is expressed as

$$\varepsilon = -n\frac{d\Phi_B}{dt} \quad \text{---} \quad (4.5).$$

The negative sign in Eq. (4.5) is due to the fact that the induced current in the coil caused by changes in the magnetic field creates an induced magnetic field that opposes the changes. Eq. (4.5) can be transformed into various forms depending on the conditions.

Maxwell equations

Just as an electric field is generated when the strength of a magnetic field changes, a magnetic field is created when the strength of an electric field changes. Maxwell[*89] summarized this principle in terms of the so-called

---

[*89] James Clerk Maxwell (6.1831 - 11.1879) was a British theoretical physicist, mathematician. He summarized electromagnetic theories such as Faraday's law and Coulomb's law dealing with electricity and magnetism and integrated into Maxwell's equations. These equations are differential ones that are the basis of electromagnetism and are highly respected as one of the greatest achievements of physics in 19th century.

Maxwell-Ampére equation in addition to the Ampére law. From this equation and Faraday's law, one can see if an alternating electric field interact with a magnetic field, an electromagnetic wave is generated.

A magnetic vector field has the two important mathematical properties with respect to the origin. These two properties form the Maxwell equations together with the two properties of the electric field. The first characteristic is the divergence of the vector field **A**[*90)], $\nabla \cdot \mathbf{A}$, which is used to determine whether **A** flows outward. The magnetic flux lines of magnetic field **B** form a closed loop without starting or ending at a point. Mathematically, the divergence of **B** is zero. That is, **B** does not diverge. This vector field is called a solenoid vector field, and forms the Gauss law of magnetic field. It is equivalent to the absence of magnetic monopoles. On the other hand, since an electric field starts and ends with an electric charge, the divergence of an electric field is not zero but proportional to the charge density. The second mathematical characteristic is the curl, denoted by $\nabla \times \mathbf{A}$, which describes whether **A** rotates. The result of the rotation is called the circulation source. The rotations of **B** and **E** are called the Ampére-Maxwell law and Faraday law, respectively. Maxwell's equations are the combination of these laws and consist of a total of four laws.

- Gauss law: $\nabla \cdot \vec{E} = \frac{\rho}{\epsilon_0}$ --- (4.6)

- Gauss law of magnetic field: $\nabla \cdot \vec{B} = 0$ ---(4.7)

[*90)] The vector field **A** related to the magnetic field **B** as $\mathbf{B} = \nabla \times \mathbf{A}$ is called magnetic vector potential.

- Ampére-Maxwell law: $\nabla \times \vec{B} = \mu_0 \vec{J} + \mu_0 \epsilon_0 \dfrac{\partial \vec{E}}{\partial t}$ --- (4.8)

- Faraday law of induction: $\nabla \times \vec{E} = -\dfrac{\partial \vec{B}}{\partial t}$ --- (4.9)

Here J is the current density and ρ is the charge density. A solid may react to an external magnetic field **B** and an electric field **E** to create its own internal magnetic and electric fields. This is hard to calculate. To overcome this problem, Maxwell's equations are modified to Eqs. (4.10) and (4.11) using the free current density J_f and the free charge density ρ_f.

- Gauss law: $\nabla \cdot \vec{D} = \rho_f$ --- (4.10)

- Ampére-Maxwell law: $\nabla \times \vec{H} = \vec{J}_f + \dfrac{\partial \vec{D}}{\partial t}$ --- (4.11)

For these equations, the magnetic and electric fields are labeled as **H** and **D**, respectively. It needs to know the charges and currents that are "trapped" in the solid, which are not as common as with the original equations.

4.2. Dielectric property of solids

A dielectric is an electrical insulator that is polarized in an external electric field. In a dielectric in an electric field, no electricity flows and only slightly moves internally from the equilibrium position to form polarization. Positive charges move in the direction of the electric field and negative charges move in the opposite direction, so that an internal electric field is formed to reduce the total electric field in the dielectric. When a dielectric consists of weakly bound molecules, the molecules not only polarize, but also rearrange, so that the axis of symmetry coincides with the electric field. Though they are insulators with respect to their low electrical conductivity, they are dielectrics with high polarizability (permittivity). Polarization means the storage of energy in dielectrics. A typical dielectric is made by placing an insulator between two metal plates as shown in Figure 47. A perfect dielectric is a material with zero electrical conductivity.[111]

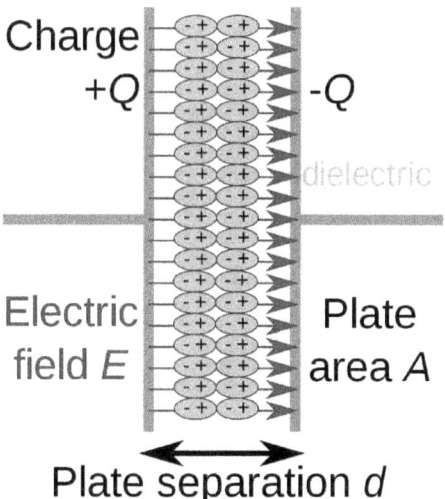

Figure 47. A representative dielectric with an insulator between two conducting plates (image from wikipedia.org).

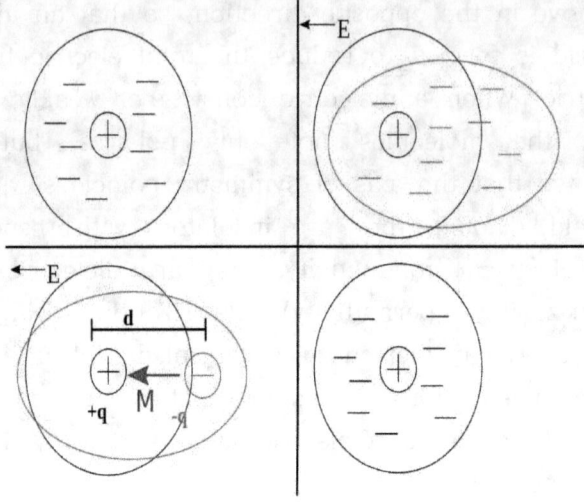

Figure 48. Interaction of an external electric field and the atom in the classical dielectric model (image from wikipedia.org).

Dielectricity

One factor characterizing the dielectric is the electric susceptibility χ_e. It is an index for a material to polarize to an external electric field **E** and is defined as

$$\vec{P} = \epsilon_0 \chi_e \vec{E} \quad \text{--- (4.12)}.$$

P is the polarization density and ε_0 is the vacuum dielectric constant. The susceptibility is related to the relative permittivity ε_r as $\chi_e = \varepsilon_r - 1$. So for vacuum χ_e is zero. The electric displacement **D** is related to **P** as

$$\vec{D} = \epsilon_0 \vec{E} + P = \epsilon_0(1+\chi_e)\vec{E} = \epsilon_0 \epsilon_r \vec{E} \quad \text{--- (4.13)}.$$

In general, dielectrics are not instantaneously polarized with respect to the external electric field, but the polarization is a function of time. This is because the relative dielectric constant varies with time.

In the classical model, materials consist of atoms, each of which is composed of a positive point charge in the center and an outer electron cloud around this charge. If there is an electric field, as shown in Figure 48, the electron cloud is distorted, forming an electric dipole. In this figure, the dipole moment is the vector denoted by **M**. The relationship between this vector and the electric field governs the behavior of a dielectric (the orientation of dipole moment is the same as that of the electric field. This is not always true, but true for many materials). When the electric field is removed, the atom returns to its original position. The time to return is called the relaxation time and the distance to the original position is exponentially attenuated.

Electric dipoles can be induced in a molecule itself (orientation polarization) or in any molecules capable of the asymmetric distortion of the nuclei (distortion polarization). Orientation polarization is caused by permanent dipoles (e.g. the oxygen and hydrogen atoms in a water molecule) and is maintained even in the absence of external electric fields. When an external electric field is applied, the orientation of polarization itself rotates without changing the distance between the negative and positive charges in each permanent dipole. This rotation does not occur instantaneously, but over time for several reasons, including the torque of the molecules. Delays in responding to the change of the external electric field produce friction and heat. When the external electric field changes at or below the frequency of the infrared region, the molecules are bent or stretched by the electric field and the molecular dipole moments change. The frequency of molecular vibration is roughly proportional to the inverse of time it takes for the molecule to bend, and this distortion polarization disappears above the infrared frequency.

On the other hand, ion polarization is a polarization caused by relative displacements between cations and anions in an ionic crystal such as NaCl. When a crystal or molecule consists of more than one sort of elements, the charge distribution around the atoms or molecules in the crystal is either positive or negative. In this case, lattice or molecular vibrations cause relative displacements of the atoms and change the positive and negative charge centers. The central location is affected by the symmetry of the displacement. If the centers do not match, polarization occurs in the molecules or crystal. This polarization is ionic polarization and causes strong dielectric effects and dipole polarization.

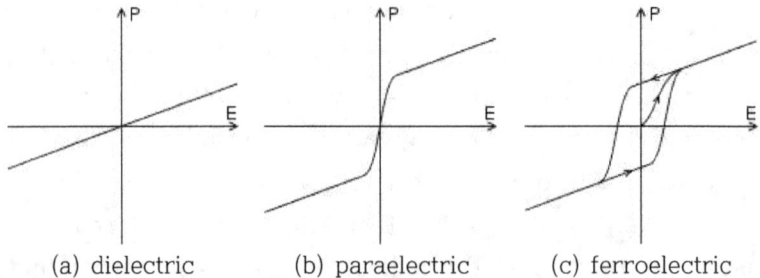

(a) dielectric   (b) paraelectric   (c) ferroelectric

Figure 49. Types of dielectric polarization (image from wikipedia.org).

Dielectrics of various kinds

If polarization occurs in a material, its moment P is almost proportional to the external electric field E. This polarization is called dielectric polarization (Figure 49a). For paraelectrics, the polarization is nonlinear.[112] The slope of the polarization curve is the permittivity, which varies with the external electric field strength (Figure 49b). Paraelectricity means that many materials (especially ceramics) polarize against external electric fields, but disappear when the external field disappears.[113] This is due to the combination of individual ion distortion (the displacement of electrons from the nucleus) and molecular or ion polarization. Paraelectricity can occur in the crystal phases in which electric dipoles do not arrange but react to external electric fields. One of the paraelectrics with a high dielectric constant is $SrTiO_3$. $LiNbO_3$ crystals are ferroelectric below 1,430 K but paraelectric above this temperature. Other perovskites*[91] are also paraelectric at high temperatures. When an electric field is applied to a paraelectric insulator, the temperature rises while polarizing, and when the electric field is removed, the temperature decreases.[114] A heat pump for cooling can be

---

*91) Perovskite is a material having the same crystal structure as calcium-titanium oxide ($CaTiO_3$). It was first discovered in the Ural Mountains and named after the Russian mineralogist L.A. Perovski (1792 - 1856). In general, the formula is represented by ABX_3, where 'A' and 'B' are cations of very different sizes and X is an anion bound to both cations. 'A' is greater than 'B'. The ideal structure is surrounded by six anions, in which B forms a octahedral structure, and A has a dodecahedron consisting of eight triangles and six squares.

configured using this principle.

Ferroelectricity is a property of any materials with spontaneous electrical polarization that can be reoriented by an external electric field.[115] Rochelle salt[*92)] is the first ferroelectric found.[116] Ferroelectrics have spontaneous polarization even at zero electric field. This spontaneous polarization reverses its orientation in a strong electric field. Therefore, the polarization is influenced by the electric field strength and the previous history, so that an hysteresis occurs as shown in Figure 49c. In general, ferroelectrics lose spontaneous polarization and become paraelectrics above the Curie temperature (T_C). Many ferroelectrics completely lose their piezoelectric properties above T_C because they have a centrosymmetric crystal structure when they become paraelectric.[117]

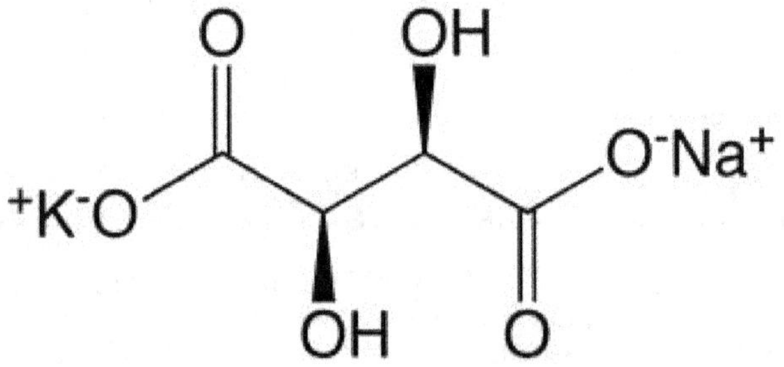

Figure 50. Molecule structure of Rochelle salt (image from wikipedia.org).

[*92)] Rochelle salt (Potassium sodium tartrate tetrahydrate, $KNaC_4H_4O_6 \cdot 4H_2O$)

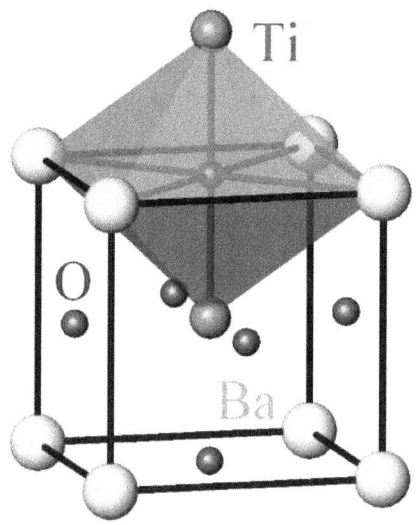

Figure 51. BaTiO$_3$ crystal structure (image from wikipedia.org).

Electric dipoles in the interiors of a ferroelectric are combined with the crystal lattice, so the dipole strength varies as the lattice deforms. Spontaneous polarization causes a change in the surface electric charge. Because of this, electricity can flow through the ferroelectric capacitor even when no voltage is applied. The crystallographic lattice structure changes due to the change of external force and temperature. Piezoelectriciy indicates the generation of electricity on the surface by external mechanical stresses. Pyroelectricity refer to the spontaneous polarization which varies with temperature.

A phase transition in ferroelectrics is due to the distortion of crystal structure, like BaTiO$_3$. Or it is an order-disorder transition as with NaNO$_2$. In BaTiO$_3$, when one ion deviates slightly from its equilibrium position, a local electric field is formed by the ions, and this force exceeds the elastic recoiling force. Thus there is a permanent dipole moment. Ionic displacement in $BaTiO_3$ is the relative positioning of titanium ions in the octahedron of oxygen. $PbTiO_3$ is crystallographically similar to $BaTiO_3$ but has more complicated interactions between the lead and oxygen ions. When a ferroelectric undergoes an order-disorder phase transition, there is a dipole moment in each unit cell, but the arrangement becomes disordered at high temperatures. Below the critical temperature, the arrangement is ordered through the phase transition.

4.3. Magnetism in solids

A magnetic field is generated by an electric current and by spin magnetic moments of elementary particles, which in turn affects the current and the magnetic moments. Ferromagnetic materials such as iron, nickel and cobalt are attracted to magnetic fields and become permanent magnets generating magnetic fields by themselves. Ferromagnetic materials are strongly influenced by external magnetic fields. Other materials also respond to magnetic fields to some extent. Paramagnetic materials, such as aluminum, are weakly attracted to external magnetic fields. In contrast, diamagnetic materials such as copper and carbon are weakly repulsive. Antiferromagnetic materials, such as chromium and spin glass,[93] are complex in response to

---

[93] In solid-state physics, spin glass is a magnet with no regular crystal structure. The magnetic spins of the atoms (the

external magnetic fields. Except ferromagnetic materials, they are generally called non-magnetic because their response is weak toward magnetism. Magnetism in materials is affected by temperature, pressure and external magnetic fields and their magnetic properties can change.

The magnetic properties of materials are mainly due to the electron magnetic moment of the constituent atoms. The amount of magnetic moment of an atomic nucleus is one-thousandth of the electron magnetic moment and can usually be ignored. However, the nuclear magnetic moment is important in nuclear magnetic resonance and magnetic resonance imaging (MRI).*94) Most of the electron magnetic moment in materials is canceled because the electrons are paired with the magnetic moments opposite to each other due to the Pauli exclusion principle. Unpaired electrons in the atoms contribute to the magnetic moment of the material, but they cannot have magnetism if their orientation is irregular. On the other hand, when the magnetic moments of unpaired electrons are arranged in

---

orientation of the N and S poles in three-dimensional space) are arranged in an orderly manner. The name "glass" comes from the similarity between the magnetic disorder in spin glass and the disorder in the atomic arrangement of materials such as window glass. In amorphous solids, the atomic bond structure is very irregular, while crystals have certain rules of interatomic bonds. In ferromagnetic materials, magnetic spins are arranged in the same direction, similar to solid crystals.

*94) An imaging technology using the nuclear magnetic resonance principle. When the high frequency from a magnetic resonance imaging device resonates with the hydrogen atom nucleus of the body, the signal from the computer is reconstructed and imaged. Unlike x-ray computed tomography (CT), it is harmless to the body. Using MRI, measuring the oxygen content of blood can provide information about blood flows in the brain.

one direction, a strong magnetic field is formed. Therefore, magnetism in a material is sensitive to how the electron magnetic moments are arranged, and it is difficult to have magnetism at high temperatures because the arrangement becomes disordered.

A magnetic field carries energy, and matter behaves in order to lower the amount of the total energy. When diamagnetic materials are in a magnetic field, the magnetic dipoles are arranged in the opposite direction to the magnetic field, reducing the strength of the total magnetic field. When ferromagnetic materials are in a magnetic field, the dipoles align in the same direction as the magnetic field, extending the walls of the magnetic domain.

Figure 52. Pyrolytic carbon has the highest negative magnetic susceptibility at room temperature. It sustains in response to the strong magnetic field of neodymium (Nd) magnet (image from wikipedia.org).

Diamagnetism

Diamagnetism is a quantum mechanical effect to oppose external magnetic fields and appears in all materials. However, diamagnetism is only observed in pure diamagnetic materials because the influence of other magnetism such as ferromagnetism and paramagnetism is strong. Pyrolytic carbon[*95)] as shown in Figure 52 is the most representative. Diamagnetic materials are commonly considered nonmagnetic materials: water, wood, organic compounds such as petroleum and plastics, metals such as copper, especially heavy metals such as mercury, gold and bismuth. In diamagnetic materials, a magnetic field is formed in the direction canceling the external magnetic field, that is, in the direction opposite to the external magnetic field.

In diamagnetic materials, there are no unpaired electrons in the atom and thus no spin-induced electron magnetic moment. In this case, magnetism is only due to the orbital motion of the electrons. All materials have these electron orbital motions, but the diamagnetism effect by the orbital motions is buried by the strong effect of unpaired electrons inducing paramagnetism and ferromagnetism.

---

*95) It is a material similar to graphite, not found in the nature. Covalent bonds exist between the graphene layers, which refers to one layer of graphite. In general, some graphite is crystallized by heating hydrocarbons to near pyrolysis temperatures. It is formed by heating artificial fibers in vacuum. It can be also produced when graphite is coated on a plate in hot gases. Due to its high heat resistance, it is used for missile heads, rocket motors, heat shields, graphite-reinforced plastics, and nuclear fuel particle coatings.

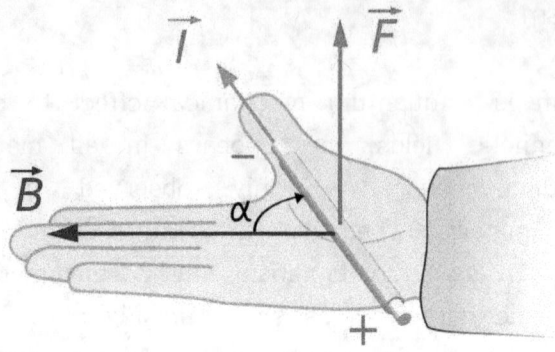

Figure 53. The Lorentz force is perpendicular to the direction of current and magnetic field. The force of a charged particle with q moving at the velocity v in the electric field **E** and the magnetic field **B** is given by Eq. (4.1) (image from wikipedia.org).

The permeability of diamagnetic materials is less than the vacuum permeability because of their repulsive behavior to external magnetic fields. In equation (4.2), **M** has a negative value. Water is a weak diamagnetic material with a relative permeability close to 1 and the susceptibility ($x = \mu-1$) being -9.05×10^{-6}, almost zero. It is -1.66×10^{-4} for bismuth and 4.00×10^{-4} for pyrolytic carbon, the absolute values of which are 1/10 of paramagnetism or ferromagnetism. All conductors exhibit diamagnetism when the magnetic field changes. The electrons are rotated by the Lorentz force (Figure 53) to form a vortex, and the vortex forms a magnetic field opposite to the external magnetic field to suppress the movement in the conductor. Very rarely, diamagnetism is greater than paramagnetism. The susceptibility of gold is less than zero and therefore it is a diamagnetic material but shows weak paramagnetism.[118]

Figure 54. A magnetic fluid showing the instability of a normal magnetic field caused by a neodymium magnet (image from wikipedia.org).

Paramagnetism

Paramagnetism means that a material is weakly attracted to an external magnetic field and a magnetic field is induced internally in the same direction as the external magnetic field. Paramagnetic materials include aluminum, oxygen, titanium and iron oxide (FeO). Most elements and some compounds are paramagnetic. The relative permeability is higher than 1, and the induced magnetic moment is linear to the external magnetic field strength. However, its strength is weak so that a quantitative measurement requires a precision device such as SQUID.*96) In general,

---

*96) (Superconducting Quantum Interference Device) It is a very precise magnetometer measuring superfine magnetic fields using superconducting technology. A SQUID can measure magnetic fields as low as 5 aT (5×10^{-18} T). Usually the magnets

the magnetic susceptibility usually ranges from 10^{-3} to 10^{-5} and synthetics such as ferrofluids (an example is shown in Figure 54) can be as high as 10^{-1}.

Paramagnetic materials have unpaired electrons. Only one electron exists in the atomic or molecular orbital. According to Pauli's exclusion principle, an electron-electron pair erases the magnetic field because the orientations of the spin magnetic moments are opposite to each other, but unpaired electrons can be arranged in any directions. A magnetic moment is formed internally along the external magnetic field so that the strength of magnetic field increases. Unpaired electrons are like tiny magnets. If there is an external magnetic field, they are arranged parallel to the magnetic field, but when there is no external magnetic field, the anisotropic orientation is lost due to thermal vibration and thus magnetism disappears. Only some electron spins can conform to the external magnetic field, and the ratio is proportional to the strength of the external magnetic field.

The atoms or molecules that make up paramagnetic materials have permanent magnetic moments even without external magnetic fields. This is, of course, due to unpaired electron spins. In pure paramagnetism, the total magnetic moment is zero because the spin array is disordered by thermal energy. In terms of classical mechanics, the magnetic dipole aligns with the external magnetic field because the torque induced by the external magnetic is the lowest when the field is parallel to the orientation of the dipole. Quantum mechanics explains it in terms of electron

---

of refrigerators have a magnetic strength of 0.01 T. Animals exhibit a very low magnetic field strength around 10^{-9} ~ 10^{-6} T.

spin and angular momentum. Ferromagnetic materials become paramagnetic above the Curie temperature[97] and antiferromagnetic materials above the Néel temperature.[98] This is because thermal vibration disturbs the arrangement of electron spins.

In modern physics, electrons move freely in a conductor, but the electrical conductivity varies from conductor to conductor, which is interpreted in relation to the electronic band structure[99] resulting from incompletely filled energy bands. In ordinary nonmagnetic conductors, the conduction band is the same regardless of the electron spin. When an external magnetic field is applied, the conduction band splits up and down by the difference in potential energy caused by the external magnetic field. The Fermi level[100]

---

[97] It is also called the Curie point. Above this temperature, a ferromagnetic material loses its permanent magnetism and becomes paramagnetic. This phenomenon was discovered by Pierre Curie (5.1859 - 4.1906). He was a French physicist who worked in areas such as crystallography, magnetism, piezoelectricity and radiation. He won the Nobel Prize in Physics in 1903 with his wife Marie Curie (11.1867 - 7.1934).

[98] A magnetic ordering temperature. An antiferromagnetic material becomes paramagnetic above this temperature.

[99] In solid-state physics, the electron band structure refers to the range of energy that the electrons may or may not have in a solid. This band structure is derived from the electron wave function in the regular crystallographic structure of a solid. The electromagnetic properties of solids, such as the electrical resistivity and optical properties, can be well explained by the band structure.

[100] It is a thermodynamic energy required to add a single electron into a solid. It does not contain the energy to bring the electron to the solid. The Fermi level is associated with the electron band structure and the electrical properties of a solid can be defined in terms of the band structure. As a hypothetical electron energy level, this level is filled with 50%

must be the same, so there is some spin movement. This is weak paramagnetism and is called the Pauli paramagnetism. This effect can be counteracted by the diamagnetism of opposite signs caused by all the paired electrons in the atom.

In general, the overlapping of the wave functions of adjacent electrons in a solid to eliminate the localization of the electrons means that the Fermi velocity (the Fermi energy expressed in terms of velocity) is high. This also means that in one band, the number of electrons is less sensitive to the change in the band energy, and a weaker magnetic field is formed. Because of this, *s*- and *p*-type metals are typically Pauli paramagnetic or diamagnetic materials like gold. In the latter case, the diamagnetism of the filled electrons overwhelms the paramagnetism of the valence electrons.

Strong magnetism typically involves the electrons in *d* or *f* orbitals. The *f* electrons are particularly localized. Lanthanide metal atoms with the atomic number of 57 to 71 can hold seven unpaired electrons, so the magnetic moment is very large. Superstrong magnets are made of these lanthanide elements, and neodymium (Nd) and samarium (Sm) magnets are typical.

The above mentioned thing is a generalization for metals that do not constitute isolated molecules. Even in the case of molecular solids, electrons can be localized. In the molecular structure, there should be no incompletely filled electron orbitals (i.e. odd spins), but there can be an unfilled shell portion within a molecule. Oxygen molecules

---

probability at thermodynamic equilibrium. The difference in the Fermi level can be measured with a voltmeter.

are a good example. Unpaired spins exist in the orbitals derived from the *p*-wave function of the oxygen atom, but only one adjacent molecule can overlap in an oxygen molecule. In the crystal structure, the distance from other oxygen atoms is too large to overcome the localization of the electron and the magnetic moment is maintained.

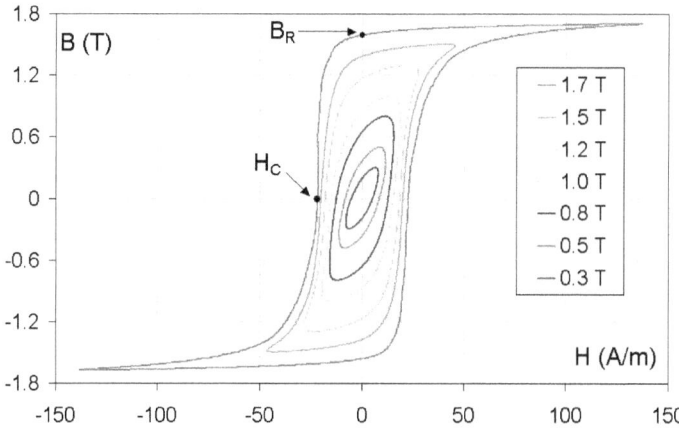

Figure 55. Hysteresis curve of an electrical steel with a preferred orientation (image from wikipedia.org). B_R is the retentivity and H_C is the coercivity.

Ferromagnetism

Ferromagnetism refers to the property of materials such as iron becoming permanent magnets or being strongly attracted to magnets. The weaker one than ferromagnetism is ferrimagnetism. Permanent magnets are ferromagnetic or ferrimagnetic materials that are magnetized by an external magnetic field and remain magnetic even when the external magnetic field disappears. What we commonly call magnets in everyday life and what are strongly attracted to these magnets are ferromagnetic materials. Refrigerator doors close well and stay in this state because of ferromagnetism. Ferromagnetic materials include iron, nickel, cobalt and alloys of these and some rare earth metals. Although the constituent elements are not ferromagnetic, some alloys become ferromagnetic.*[101] In some cases, though all of the constituent elements are ferromagnetic, like stainless steels, they are nonmagnetic.

Rapid cooling of liquid alloys can lead to the formation of amorphous ferromagnetic alloys. They are advantageous because magnetism is almost uniform and isotropic. This alloy has a low coercivity*[102] and hysteresis loss, high permeability and electrical resistivity. Transition metal-metalloid alloys are typical of this kind.*[103] Exceptionally strong ferromagnetic materials are rare earth magnets. These magnets, which contain lanthanides, are known to have strongly interacting magnetic moments localized in the f-orbital. Most ferromagnetic materials are metals because conduction electrons are often involved in the ferromagnetic interaction.

Actinide-based alloys are ferromagnetic at room temperature or become ferromagnetic when cooled. PuP is paramagnetic at room temperature, but when cooled below

---

*[101] Found by German mining engineer and chemist Fritz Heusler (2.1866 - 10.1947), it is called the Heusler alloy.

*[102] When a material becomes magnetized under the influence of an external magnetic field, the magnetic field is maintained to some extent even if the external magnetic field is disappeared. See the hysteresis curve in Figure 55.

*[103] Alloy of 80% transition metals (usually Fe, Co, Ni) and metalloids with lower melting points (B, C, Si, P, Al, etc.).

the Curie temperature (125 K), it becomes ferromagnetic accompanied by the phase transition from the cubic crystal structure to the tetragonal one. In the ferromagnetic state, the preferred orientation of PuP is <100>.[119] $NpFe_2$[120] with the preferred orientation of <111> is paramagnetic and cubic above ~500 K. Cooling below the Curie temperature distorts the angle of the rhombohedral structure from 60° (cubic) to 60.53°. Similar lattice distortion occurs in $NpNi_2$ below 32 K. The strain is about $43×10^{-4}$.[121] In 2009, the physicists of the Massachusetts Institute of Technology demonstrated that ferromagnetism appeared when lithium gas cooled*[104] below 1 K.[122] It was the first demonstration that ferromagnetism also appears in a gas.

The electron has a charge and also has a magnetic moment in itself. The dipole moment is due to the electron's quantum mechanical spin. The spin is either "up" or "down" depending on the orientation of the magnetic field. This is the main source of ferromagnetism and partly due to the orbital angular momentum around the nucleus. When these small electron magnets are arranged in a direction, each magnetic force add up. When the electron shell is filled, the total magnetic moment is zero because the spins are opposite in electron pairs. If the electron shell is not completely filled and there are unpaired electrons, the total moment is not zero. Therefore, ferromagnetism appears in materials with many unpaired electrons in the atoms. These unpaired magnetic dipoles are aligned along the external magnetic field. This is paramagnetism. Ferromagnetism refers to the magnetization

---

*[104] Li-6 was cooled to 150 nK or less using an infrared laser.

due to the spontaneous arrangement of the dipoles even without external magnetic fields. Ferromagnetic materials, like paramagnetic ones, contain unpaired electrons, and their magnetic moments are aligned in the same direction, resulting in a lower energy state. Even in the absence of an external magnetic field, the magnetic moments of these electrons spontaneously align in the same direction. Ferromagnetic materials lose their ferromagnetism when the magnetic arrangement is disordered due to thermal energy above the Curie temperature.

When two adjacent atoms have unpaired electrons, the exchange interaction varies depending on whether the two electron spins are parallel or antiparallel. This affects the electron position and the coulomb attraction, resulting in a energy difference in these two states. The exchange interaction is due to the Pauli exclusion principle that two electrons with the same spin cannot have the same quantum number. Thus, under some conditions, when the orbitals of unpaired electrons overlap with those adjacent atoms, the charges are farther apart than when the spins are parallel. Therefore, the electrostatic energy of the electrons with parallel spins is lower and more stable than the opposite case. This energy difference is called the exchange energy, and can be greater than that due to the dipole-dipole interaction (which causes the dipoles to align antiparallel) along the orientation of magnetic dipole.[123] For ferromagnetic materials the exchange interaction is considerably (for iron 1,000 times) stronger than the dipole-dipole interaction. Therefore, below the Curie temperature, the dipoles of ferromagnetic materials basically have the same arrangement.

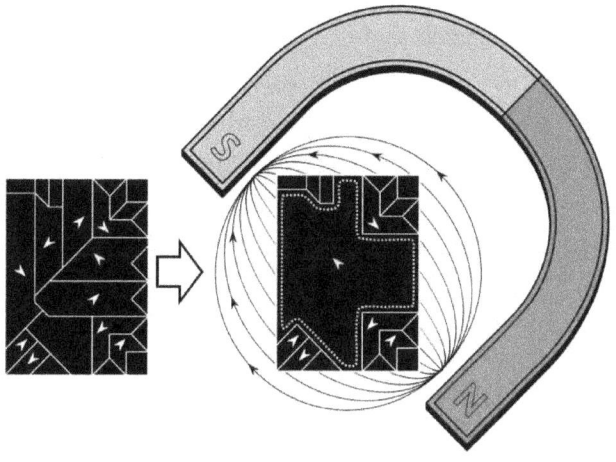

Figure 56. Magnetic domain boundaries (white lines) and the effect of a magnetic field on the domain structure (modified image from wikipedia.org).

Every atom in ferromagnetic materials is like a small permanent magnet, and all the magnetic moments of the atoms in a zone are arranged in the same direction. This zone is called a magnetic zone or Weiss*[105] domain. This magnetic zone can be seen with a magnetic microscope. In the absence of external magnetic fields, if the magnetic zone is too wide, it becomes unstable and is divided into two zones in the opposite and stable orientations. Applying an external magnetic field moves the zone boundary and increases the zone area in the direction of the external magnetic field. This state can be maintained even if the

---

*[105) Pierre-Ernest Weiss (3.1865 - 10.1940) was a French physicist. In 1907 he developed a theory of the magnetic domain of ferromagnetism.

external magnetic field is removed. This process is called magnetization, and permanent magnets are formed through this process. When the magnetization is so strong that all the zones merge into one, the magnetism saturates. When a magnetized material is heated above the Curie temperature, thermal vibration causes the domain structure to be broken down and magnetism is lost. On the contrary, the domain structure is immediately restored when cooed, as a liquid becomes a solid.

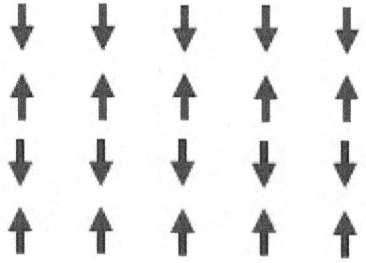

Figure 57. Array of magnetic moments for antiferromagnetism

Antiferromagnetism

Unlike ferromagnetic materials, inherent magnetic moments of unpaired electrons are arranged in the opposite direction in antiferromagnetic materials. It is called antiferromagnetism when the magnetic moment of one atom is arranged opposite to the magnetic moment of the neighboring atom as shown in Figure 57. Antiferromagnetic materials have zero magnetic strength and do not form a magnetic field. In the presence of an external magnetic field, the absolute value of the magnetic strength of a sub-lattice may be different from that of other sub-lattices, resulting in a non-zero net magnetic moment even in antiferromagnetism.

At zero absolute temperature, the magnetic strength should be zero, but weak magnetism is formed by the spin-canting effect seen in hematite. Spin canting is resulted from the combination of two opposite causes. While the spins of the two electrons should be opposite to each other, the spins become 90° to each other due to the antisymmetric exchange effect caused by the relativistic effect of the spin-orbital combination. As a result, the spin axis tilts relatively as shown in Figure 58.[124]

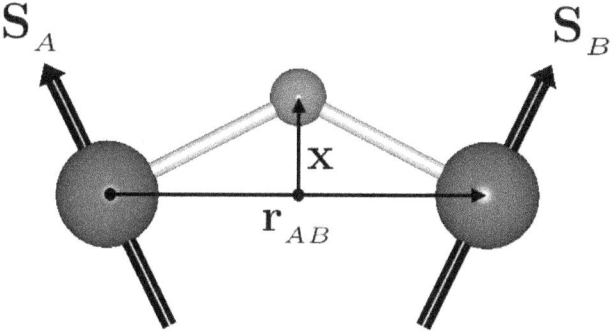

Figure 58. Spin canting: the electron spin in hematite is tilted (image from wikipedia.org).

Like other types of magnetism, the formation of antiferromagnetic structures is the result of exchange interaction between the magnetic moment and spin. Interactions occur between adjacent spins in a simple cubic lattice. Materials are divided into ferromagnetism or antiferromagnetism according to the interaction, which may induce a lattice deformation (namely the geometrical frustration) to form a complex structure that is neither ferromagnetism nor antiferromagnetism. There are usually many antiferromagnetic materials in transition metal compounds, especially oxides.*106) Antiferromagnetic materials can be combined with ferromagnetic materials by forming a ferromagnetic layer on the surface, in which case the orientation of the ferromagnetic material is fixed and used for magnetic sensors.

Unlike ferromagnetism, several ground states may exist in antiferromagnetism. In one dimension, the ground state of diamagnetism is a state in which the spin alternates up and down, but there may be multiple ground states in a two-dimensional structure. Of the three electron spins at the vertices of a equilateral triangle, one electron spin is "up" or "down" and the total possible states are 2^3 = 8, out of which the number of ground states is 6. The remained two states are with all the spin "up" or "down". Such multiple ground states are found in Kagome lattices (Figure 59, $Co_3V_2O_8$)[125] or hexagonal lattices.

Artificial antiferromagnetic materials can be made by stacking two or more ferromagnetic thin layers separated by nonmagnetic thin layers.[126] The combination of dipole moments of ferromagnetic layers forms an antiparallel array of ferromagnetic materials.

In some cases, such as iron phosphate ($FePO_4$) glass, amorphous materials become antiferromagnetic below the

---

*106) Hematite, metals like chromium, alloys like FeMn, oxides like NiO.

Néel temperature. This disordered structure destroys the antiparallel state of the adjacent spins. In other words, it is impossible to form a three-dimensional structure in which individual spins are surrounded by the opposite spins. Antiferromagnetism is determined only from the averaged relationship with the adjacent spins. This magnetism is also called spero magnetism.

Antiferromagnetic materials are rare and are usually observed at low temperatures. Depending on the temperature change, antiferromagnetic materials become diamagnetic and ferromagnetic. In some materials, neighboring pairs of electrons are arranged in the opposite directions, but there is no preferred geometry. This is called spin glass and is a kind of geometrical frustration.

Figure 59, Kagome lattice pattern. The spin array of magnetic ions in $Co_3V_2O_8$ has a Kagome lattice pattern and has excellent magnetic properties at low temperatures (image from wikipedia.org).

Ferrimagnetism

Like ferromagnetic materials, ferrimagnetic materials are magnetic even without external magnetic fields. However, like antiferromagnetic materials, the spins of neighboring electron pairs are opposite, as shown in Fig. 60. As with Fe^{2+} and Fe^{3+} ions, magnetism occurs because the sum of the magnetic moment in one orientation is different in absolute value from the other orientations.[127] Most ferrites (Fe_2O_3 or Fe_2O_3 based magnets) are ferrimagnetic. Magnetite (Fe_3O_4) is also a ferrimagnetic, not a ferromagnetic material. Before Louis Néel*[107] discovered ferrimagnetism and antiferromagnetism in 1948, it was known as a ferromagnetic material.[128] Ferrimagnetic materials include YIG, hexagonal ferrite ($PbFe_{12}O_{19}$ or $BaFe_{12}O_{19}$) containing iron oxide and other metal elements, and $Fe_{1-x}S$ (pyrrhotite).[129]

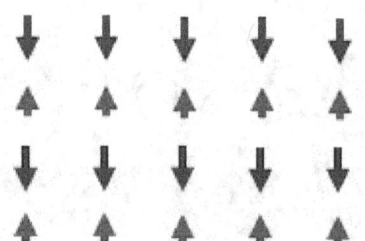

Figure 60. A ferrimagnetic array

---

*107) Louis Néel (11.1904 – 11.2000) was a French physicist. In 1970 he won the Nobel Prize in Physics for the study of magnetism in solids.

Ferrimagnetic materials, like ferromagnetic ones, are magnetic below the Curie temperature and paramagnetic above it. However, sometimes the sum of each opposite dipole moments is the same below the Curie temperature, and magnetism disappears. This temperature is called the magnetization compensation point (see Figure 61). This is easily observed in garnets and rare earth transition metals such as YIG. Furthermore, there is an angular momentum compensation point where the total amount of angular momentum is zero. This is important for clearing magnetization at a high speed in magnetic storage devices.[130]

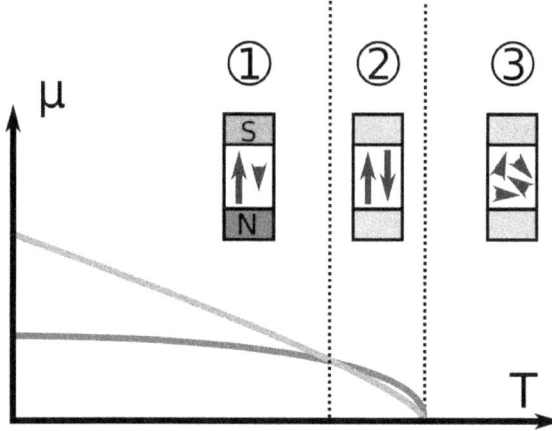

Figure 61. ① Below the magnetization compensation point, a ferrimagnetic material creates a magnetic field. ② At the compensation point, the magnetic moments cancel each other out and the total moment is zero. ③ Loss of magnetism above the Curie temperature (image from wikipedia.org).

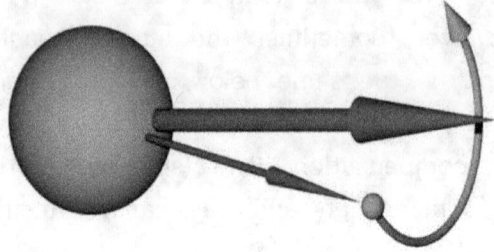

Figure 62. Precession of negatively charged particles. The large arrow is the external magnetic field and the small arrow is the orientation of the spin angular momentum of the particle (image from wikipedia.org).

Ferrimagnetic materials are highly anisotropic in resistivity and magnetism. The magnetic anisotropy is induced by an external magnetic field. When the external magnetic field coincides with the magnetic dipoles, the total magnetic dipole moment is not zero and the magnetic dipoles precess at the frequency controlled by the external magnetic field. That is, the axis of rotation of the magnetic dipole changes. This frequency is referred to as the Larmor[108] frequency or precession frequency. For example, as the precession interacts strongly with the magnetic dipole moment, the microwave signal is circularly polarized in the same direction. If polarization occurs in the opposite direction, the interaction becomes minimal. If the

---

[108] Sir Joseph Larmor (7.1857 - 5.1942) was a Irish physicist and mathematician. He is a pioneer in electricity, kinetics, thermodynamics, and electronic theory of matter.

interaction is strong, microwave signals penetrate the material. Using these properties, various microwave devices, isolators,[*109)] circulators,[*110)] and gyrators can be manufactured.

4.4. Theory of magnetism

A magnetic dipole is formed by a permanent magnet or by a current flowing through a closed circuit, as shown in Figure 63. The magnetic south and north poles caused by Earth's magnetic field is an example of a magnetic dipole. Earth's north pole (now it is in the Arctic Ocean, north of Canada) is physically the south pole. This is because it is attracted to the north pole of a compass. In a strict sense, a permanent magnet or Earth is not a magnetic dipole, but a collection of magnetic dipoles. Is it possible to separate the negative and the positive magnetism like electricity? Since a bar magnet is made of ferromagnetic materials formed by the uniform distribution of electrons, cutting the bar magnet in half results in smaller bar magnets. There are N and S poles, but they cannot be separated. If there is a monopole, it is essentially a new magnetic material.

[*109)] It is a two-port device that carries microwaves or radio waves in only one direction. This protects the equipment by blocking harmful microwaves.
[*110)] A device with three to four ports. The device rotates and sends out incoming microwaves or radio waves.

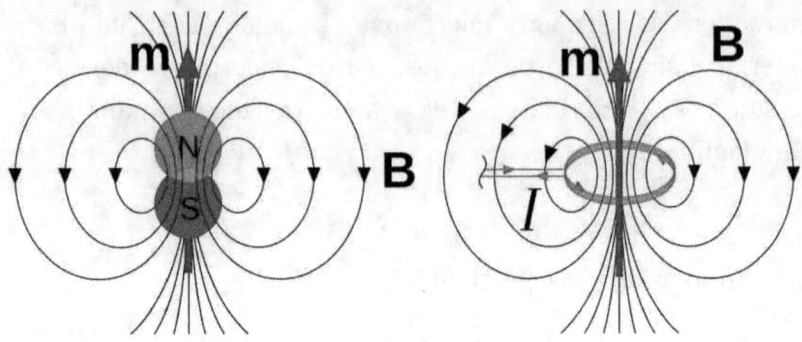

Figure 63. Magnetic field **B** and magnetic moment **m** of a permanent magnet (left), **B** and **m** by a current flowing in a closed loop (right) (image from wikipedia.org).

Magnetic monopole

A magnetic monopole has a so-called "magnetic charge". In the grand unified theory (super-string theory), monopoles exist,[131] but no magnetic charge such as electric charges has been found.[132] In some solids quasi particles that look like magnetic monopoles have been observed,[133] and magnetic monopoles have been found only mathematically.[134]

Maxwell's equations are not symmetric in terms of electric and magnetic fields. With magnetic charges, the equations can be completely symmetric. From these imaginary magnetic charges, the "magnetic current density, j_m" can be introduced and Gauss's law of equation (4.7) is then modified as

$$\nabla \cdot B = j_m \quad \text{--- (4.14)}.$$

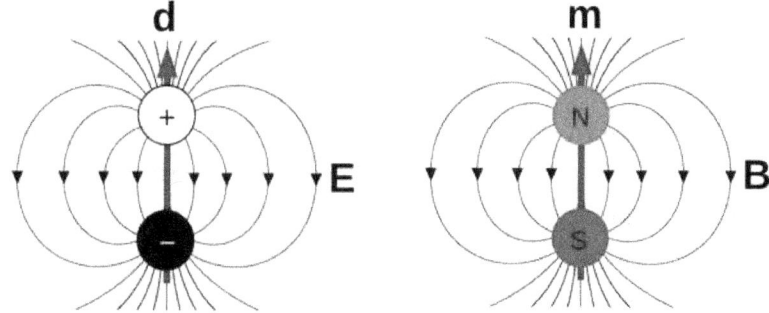

Figure 64. The magnetic field due to a magnetic monopole (right) similar to the electric field (left) by a charged elementary particle (image from wikipedia.org).

Electricity and magnetism have some symmetry, and quantum theory can distinguish individual negative or positive magnetic charges, so the S and N poles can be imagined as shown in Figure 64. In 1931 Dirac proposed the presence of magnetic monopoles, and stated, if quantized electric charge are present, the magnetic charge of the magnetic monopole would also be quantized.[135]

The grand unification theory also predicts magnetic monopoles. As quantum field theory and gauge theory*[111]

---

*111) Field theories dealing with electric and gravitational fields usually refer to the dynamics of field and define the changes by other independent physical variables that affect the fields, including time. Usually these fields are described as Lagrangian or Hamiltonian equations and handled as classical or quantum mechanical systems with an infinite number of degrees of freedom. Gauge theory is a kind of field theory, a specific mathematical form that handles many degrees of freedom of an equation called Lagrangian. The transformation between gauges forms a Lie group, which is called the symmetry group or gauge group. The gauge field belongs to Lagrangian and does not change in a local group transformation. This is called

integrate electromagnetism and the weak interaction, and as the theory of strong interaction develops, they have been unified and into a theory. This is the grand unification theory. This theory assumes particle called dyons that allow magnetic monopoles. Dyons are imaginary particles with both electric and magnetic charges and are stable because they do not decay into other particles. Early cosmological models predicted a tremendous density of magnetic monopoles, which differed greatly from the experimental results.136 If there were a great number of monopoles, the expansion of the universe were impossible long time ago. In current theory, the monopole itself is not denied, and most of the monopoles are assumed to be extinguished by the initial cosmic expansion and are now considered almost impossible to observe.137

Quantum mechanical origin of magnetism

There are things that are not explained by classical mechanics in diamagnetism, paramagnetism, and ferromagnetism. These can be fully explained by quantum theory.138 In 1927 Heitler*112) and London published a theory of quantum mechanics in which a hydrogen

---

gauge invariance. In mathematics, Lie groups are a contiguous (differential) set in which each element is described by several existing variables. Thus, in Li groups a natural model of the concept of continuous symmetry, such as rotational symmetry in three dimensions can be obtained.

*112) Walter Heinrich Heitler (1.1904 - 11.1981) was a German physicist. He contributed to the development of quantum electrodynamics and quantum field theory. The theory of electrons involved in chemical bonding led chemistry to quantum mechanics.

molecule is formed from orbitals u_A and u_B centered on the nuclei of the two hydrogen atoms A and B. According to this theory, a σ-orbital is formed between two molecules.

$$\psi(r_1,r_2) = \frac{1}{\sqrt{2}}(u_A(r_1)u_B(r_2) + u_B(r_1)u_A(r_2)) \quad \text{---} \quad (4.15).$$

The first electron r_1 is a hydrogen-orbital centered at the second nucleus, while the second electron r_2 rotates around the first nucleus. This "exchange" phenomenon is an expression of the quantum mechanical property that particles with the same properties are indistinguishable from one another. This applies to magnetism as well as chemical bonds. This exchange is the core of magnetism and is 100 times stronger than the electromagnetic dipole-dipole interaction energy.

In the spin function $\chi(s_1, s_2)$, which is the origin of magnetism, the symmetry function (+ sign) is multiplied by the antisymmetric function (- sign) according to Pauli's exclusion principle. Therefore,

$$\chi(s_1,s_2) = \frac{1}{\sqrt{2}}(\alpha(s_1)\beta(s_2) - \beta(s_1)\alpha(s_2)) \quad \text{---} \quad (4.16).$$

That is, not only u_A and u_B are replaced by α and β, respectively, but the sign is replaced by + to - and r_i is replaced by the specific value s_i (= ± ½). Thus α(+ ½) = β (-½) = 1, α(-½) = β(+ ½) = 0. The "single state" or "- sign" means that the spins are antiparallel, yielding antiferromagnetism, and that molecules made of two atoms become diamagnetic. The tendency to chemical bonding (the formation of symmetric molecular orbitals by the + sign) automatically leads to the antisymmetric spin state (- sign) through the Pauli principle.

On the contrary, the coulomb repulsion, i.e. the tendency of electrons to repel one another, makes the two particles to be an antisymmetric orbital function (- sign) and complementarily causes a symmetric spin function (one of the so-called "triple functions"). So the spins will be parallel to each other. Examples are ferromagnetism in solids and paramagnetism in a gas molecular made of two atoms. With the exception of ferromagnetic materials such as iron, cobalt, nickel and some rare earths, most metals are dominated by the first mentioned tendency and become nonmagnetic (Na, Al, Mg, etc.) or antiferromagnetic (Mn). A gas molecule made of two atoms is almost diamagnetic and not paramagnetic. However, oxygen molecules are an exception because they involve the π-orbital as shown in Figure 65.

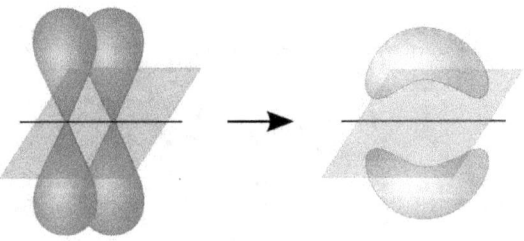

Figure 65. The two *p* orbitals of an oxygen molecule combine to form a π-orbital (image from wikipedia.org).

V. New paradigm of electromagnetism

Is an electric current a flow of electrons? This question is the topic of this book. Then the question, what is the electron?, will be another topic. Every electron has its own spin with the same amount. It is evident that this is a spin, because the magnetic field formed by this spin is fundamentally identical in magnitude to the magnetic field produced when electricity flows in a closed circuit. If the magnetic field generated by the electricity flowing through a closed circuit is due to the flow of electrons, how does the electron spin generates a magnetic field when there is no electron flow? Here we see that a magnetic field can be generated without the flow of electrons. The wave-shaped electromagnetic field in light has no electrons involved. Then it can be assumed that electrons may not be involved in the electric current in solids. In Chapter 5, we set up a new paradigm of electricity and magnetism from the point of view that the electricity in solids is not the flow of electrons but the propagation of some interatomic deformation states carrying electrical energy.

5.1. Electric current in solids

Point particle - electron

In current physics, the movement of electrons in a solid is the electrical current, and this current is a well-established one that has been dealt with for a long time as a representative electric phenomenon of solids. However, it has not been observed that electrons actually move in the form of particles in the interiors of a conductor. There appears no way to directly observe them. The conduction phenomenon is explained by the free electron model in which that some valence electrons in a metal conductor move freely and when they collide with metallic ions, their freedom is restricted to yield an electrical resistance.[139]

Electrons were identified as negatively charged particles in high vacuum cathode tube tests in the late 1800s. Then, in 1900, it was confirmed that beta rays emitted from radium isotopes were bent in electric fields and the charge/mass ratio was found to ge the same as that of the cathode tube beam. The electron was thus recognized as a component of the atom. However, these electrons are point particles, which have no volume.[140] Therefore, it is difficult to be thought of any substance. Electrons from the cathode experiments or beta-ray electrons would be any blurred energy lumps formed by the distortion of the solid vacuum in the new vacuum paradigm. In 1927, G.P. Thomson*[113]

---

*113) George Paget Thomson (5.1892 - 9.1975) was a British physicist who discovered the interaction of electrons in crystals. This proved the wave characteristics of electrons, for which he won the Nobel Prize in Physics in 1937.

collided these energy lumps with a very thin metal film and observed diffraction patterns indicating the wave nature of electron.[141] The electron is a kind of wave rather than any particle. This is the same concept as matter wave discussed in Chapter 2 of the book, Origin of Gravity and New Cosmos (ISBN 9781713042020). The flow of electricity is not the flow of electrons, but the transfer of some deformation energy stored in the solid vacuum. There are many phenomena that are not currently interpreted under the concept of charged particles in solids. Among them, superconductivity, especially high temperature superconductivity, is not well explained. In addition, the difference in the electrical conductivity of individual elements is not explained in detail.

Various forms of electron

Electrons exist in a variety of states. Free electrons moving through the vacuum, electrons bound to atoms, electrons moving freely in conductors, etc. Are these electrons all the same?

Let's first consider the electrons traveling through the vacuum. The electron has a mass but no volume, as light has no volume. Electromagnetic waves also have energy, and according to Einstein's mass-energy equivalence, light has a mass. This means that there is energy to distort the solid vacuum. The electron is a point particle with quantized energy. As a result of vacuum polarization electrons may be generated from light in the vacuum.*[114] If

---

*114) In the introduction of the photoelectric effect, we discussed the possibility that when light is irradiated to a conductor, an electron is generated from a light wave (it is actually a transformation of an electromagnetic wave), the light energy will not propagate and will be concentrated at a point in the form of a circular standing wave shown in Figure 3. In other words, the electron in the vacuum is a special form of light. Like light, the movement of electron should be wave-like. This is supported by the interference of electrons discovered by G.P. Thompson.

It was discussed in Chapter 2 that when a free electron is captured by a proton, they form a hydrogen atom and when the proton and electron fuse, they become a neutron. In the case of a circular standing wave, the direction of rotation can be divided into clockwise and counterclockwise, which may be the origin of electron spin. If one electron is generated via vacuum polarization, one positron should be also generated. Where will the positron go? If the vacuum is an infinite ocean filled with negative energy particles like Dirac's model,[115] and the negative particles form regularly ordered lattice points, the positron will be easily captured and protonated. If the lattice point has a negative charge, it cannot capture an electron in terms of the electromagnetic interaction. When the lattice point is protonated, an electron may be captured by the proton and kept stable by lowering the electromagnetic energy. But the electron still has enough energy (they lose some of their free energy), enveloping the proton in the

---

electrons are not emitted from the interiors of the conductor but are made from light on the surface. See Section 5.4 Photoelectric Effects.

[115] See section 1.3. The birth of electron and positron. Our vacuum model assumes that the solid vacuum is a dense ordered lattice like an fcc lattice.

form of electromagnetic wave. In this case, the ionization energy of a hydrogen atom is 13.6 eV, which is the potential energy of the electron with respect to the proton.

Next, let's consider the electrons in atoms. Of particular note, except for hydrogen, the nucleus consists only of protons and neutrons, in which the protons and neutrons usually are in a 1:1 ratio, with some neutron excess (see Figure 19). In the new paradigm, an atom (or materials made of atoms) have a specific volume, which exerts a stress to the surrounding solid vacuum and also back to the atom itself. At the center of a spherical object such as the stellar center there forms an singularity with an infinite compressive stress. To mitigate this situation, we imagine that a massive element has an internal structure composed of excess neutrons or filled with the cold solid vacuum only in the central region, as shown in Figure 20. Except for these excess neutrons,[116] all elements except protium (^1H) are very stable when electron-proton-neutrons (called PeNs) are the elementary building blocks of the atom and a pair of PeNs forms a stable ^4He unit. The state of the electron, proton and neutron in a PeN is not fixed but will constantly be changing in "dynamic equilibrium"[117] as in

---

[116] In the new vacuum paradigm, the excess neutrons are mainly in the center of the atom and are in a very stable state where energy is exhausted.

[117] A dynamic equilibrium occurs when there is a reversible reaction in chemistry and physics. The materials before and after the reaction are constantly changing, but there are no apparent changes because the rate of the forward and that of the reverse reaction are the same. This is also called the steady state. The forward and reverse reactions continue to occur, but the reaction appears to have stopped because the outputs of both reactions are exactly offset.

Eq. (2.1)

Finally, the electrons flowing through a conductor are considered. We have already stated that the electrons flowing through a conductor cannot be identified. While the electrons traveling in the vacuum can be observed in some way, there is no way to observe the electrons in a conductor. Although an electric current is explained by the movement of free electrons in the conductor, it is impossible to explain unusual physical phenomena such as superconductivity. Quantum mechanics explains the flow of electricity in terms of the wave functions of free electrons. There is no particle called 'electron' in the regime of the new atomic model. If the flow of electricity in a conductor is not the flow of electrons, how do we understand this flow? Should it be interpreted as the movement of negative and positive ions as with the electrical conductivity of electrolytes? We may find an answer by observing the mode of the light propagation. The current can be interpreted as the movement of electric field. In this case, we do not concern about the free electron movement. A local electric field can be formed according to the relative positioning of valence PeNs*[118] of the atoms in solids. This local field is an excited state with high potential energy, and the electric current can be understood as the movement of this excited state. In the new paradigm, this exited state is not caused by the electron itself, but is due to the distortion or deformation of a secondary bond between a pair of valence PeNs of the two adjacent atoms under the influence of an external electric field and the current is the propagation of

---

*118) A valence PeNs is a valence electron in the conventional atomic model.

the excited state along the external field.

5.2. Non-electron model of conductivity

Movement of the distortion - electric current

We have proposed a new atomic model in Chaper 2, for which plectons (proton-electron pairs) instead of separated electrons and protons within an atom were introduced. The electron is inseparable from the proton in ordinary matter. Hence, in this chapter it is assumed that the electric current in a conductor is not the flow of free electrons, but the flow of potential energy of the constituent atoms, and comprehensively the movement of information for the specific locations of valence PeNs on the surface of the atoms. That is, there are no electrons moving freely to transfer electricity, and the transfer of potential energy arising from the relative locations of valence PeNs of the atom with respect to those of neighboring atoms is the electric current in solids. In this process, the charge and discharge of potential energy occurs continuously and repeatedly in one direction along the external electric field. This process appears to be similar to the propagation of light in the solid vacuum. However, unlike the movement of light, it takes time for the valence PeNs participating in secondary chemical bonds to be rearranged (distorted) in the conductor to a higher energy state. It is similar to the case of a watermill in Figure 66. It takes time for each compartment to be filled with water and for the watermill to overcome the resistance of its rotational axis. An external electric field resembles the potential energy of water in the watermill. Just as several compartments participate in the rotation of the watermill continuously, electricity is continuously flowing when plural number of secondary interatomic PeN bonds are involved.

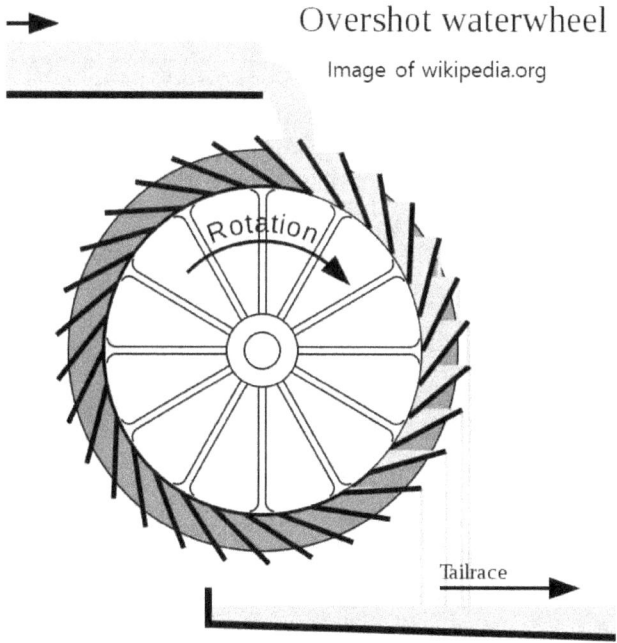

Figure 66. When water fills the compartments of the watermill, it gets power to turn the watermill. The mill shaft works to transfer energy or loses energy due to friction. The continuous process of this process resembles an electric current.

Table 4. Periodic Table of the electrical conductivity
(MS/m or Mm/Ω) @20°C and 0°C

H																	He
Li 10.7 11.8	Be 30.3 18											B	C	N	O	F	Ne
Na 21.2 23	Mg 23.8 25											Al 37.4 40	Si 9.6	P	S	Cl	Ar
K 14.7 15.9	Ca 27.0 23.5	Sc 1.5	Ti 1.8 1.2	V 0.5	Cr 7.5 6.5	Mn 0.6	Fe 9.9 11.2	Co 15.7 16	Ni 14.5 16	Cu 59.0 64.5	Zn 16.8 18.1	Ga	Ge	As	Se 8.3	Br	Kr
Rb 8.2	Sr 4.4	Y 1.8	Zr 2.2	Nb 6.2	Mo 17.5	Tc	Ru 13.0	Rh 21.2	Pd 9.2	Ag 61.4 66.7	Cd 13.7 15	In 11.3	Sn 7.9 10	Sb 2.4	Te	I	Xe
Cs 5.0	Ba 2.0 1.7		Hf 3.1	Ta 7.4	W 18.5	Re 5.3	Os 11.4	Ir 19.6	Pt 9.4	Au 45.5 49	Hg 1.0	Tl 6.0	Pb 4.8 5.2	Bi 0.9 1.0	Po	At	Rn

Transfer of electrical energy by secondary bonds

Valence PeN-based chemical bonds exist in various forms. Ionic and covalent bonds discussed in Chapter 3 are types of primary PeN bonds, and hydrogen bonds and van der Waals forces can be regarded as secondary PeN bonds. Metallic bonds are regarded as a type of primary bonds in the sense that valence PeN determines the crystal structure, but the bonds can also be secondary bonds which are responsible for the electromagnetic properties. Primary bonds are usually rigid and determine the crystal structure of solids, and secondary bonds affect the electromagnetic properties of solids. Hydrogen bonds would be regarded as a kind of primary bonds because they have some features of covalent bonds, as mentioned in Section 3.3, and therefore have a significant effect on the crystal structure of solids as with ice.

Since the electrical conductivity or resistivity of materials is dependent on temperature, we consider these properties in terms of the temperature-dependent density of secondary bonds. One of the electrostatic interactions of the van der Waals forces, the Keesom force, is an example of secondary bonds. Magnetic arrangements (e.g. phase transitions from paramagnetism to ferromagnetism and antiferromagnetism) due to the unpaired electron spins in the atoms are also highly temperature sensitive, and thus are also regarded as a type of secondary PeN bonds. As will be discussed in Chapter 6, secondary bonds are involved in the change of heat capacity of superconductors near the critical temperature, as shown in Figure 90, In general, when electricity flows in response to an external electric field in a conductor, the crystal structure based on the primary bonds does not change. However, if the primary bonds making the solid crystal structure are distorted by external forces, the secondary bonds will also be affected. In piezoelectric elements the crystal structure is distorted by external forces and electricity flows. On the contrary, when an external electric field is applied, the primary bonds in the solid can be distorted.[142] Therefore, we can state that primary bonds and secondary bonds are not distinguished clearly with regard to the electrical conductivity.

In the conduction model based on the new atomic model free electrons in the conductor are replaced by a pair of valence PeNs of two adjacent atoms, making a secondary interatomic bond. But primary bonds are not completely immune to external electric fields. They also influence the flow of electricity to some extent. Therefore, in the new model of conductivity, part of the primary bonds that react to external electric fields are also regarded as secondary bonds, and secondary bonds represent all types of interatomic PeN bonds that interact with electric fields in this book. We call them specifically electromagnetic secondary bonds, in short ε_P bonds throughout the rest of this book.

ε_P bonds at a given temperature are thermally vibrating and the individual PeNs in the bonds occupy the equilibrium positions. However, when an external electric field is applied, the thermal equilibrium is disturbed to form a new equilibrium via relocating the individual PeNs. ε_P bonds arrange along the external electric field, though they are still thermally vibrating. A deformed ε_P bond is an excited one storing electrical energy. Two or more valence PeNs, as with transition metals, may be involved in the bond and the electrical conductivity of a solid will vary depending on whether the bond deformation is easy. However, if the deformation is confined to only two adjacent atoms, no electrical conduction will occur. This is the same situation as a dielectric in Figure 47. In this case, the electrical energy is locally stored but not transferred.

The movement of one valence PeN belonging to an atom (or molecule) in a solid to a new location in respond to the external electric field also affects the locations of valence PeNs of the other atoms, which in turn will have a chain effect on the whole atoms in the solid. If this displacement of PeNs occurs immediately (elastically), electricity flows without resistance. However, as shown in the previous example of the dielectric (see Figure 49), the displacement by the external electric field takes time for activation (it is a viscoelastic reaction), resulting in electrical resistance, which will appear mostly as frictional heat in the solid (similar to the axis friction of the watermill in Figure 66) and the remained non-thermal energy goes to the final destination. Consequently, the flow of electricity is an electrically viscoelastic behavior of solids. The electrical conductivity will be determined by the number of PeNs involved, the activation energy and the time for the rearrangement of valence PeNs on the atom, which is also related to the specific structure of ε_P bond. Conductors are low in the activation energy, while the activation energy is high for insulators, and intermediate for semiconductors. In conductors, if the number of ε_P bonds to be activated is large, the electrical conductivity may be high or low depending the specific crystallographic structure of solids.

Thermal and electrical energy

Conventional theories on electrical conductivity have been established in terms of the movement of free electrons in a conductor as the flow of electricity.[143] However, the flow of electrons in the conductor has not been actually observed. The internal flow of electrons is accepted to be real because the predictions by the free electron model are consistent with the directly measurable phenomena, such as electrical resistance. There may be more than one theory that actually yields the same results, and current theories cannot explain superconductivity. Theories of normal conductivity and superconductivity are different. In the regime of the new paradigm, the two conduction phenomena are interpreted principally as the same. Free electrons in the theory of normal conductivity are replaced by the movement of ε_P bonds. Without an external electric or a magnetic field, the number and structure of ε_P bonds in a conductor are in thermal equilibrium. This equilibrium may be maintained due to van der Waals forces in Section 3.4 or in any energetically stable arrangements (or couplings) of valence PeNs, as shown in the examples of ferroelectric and ferromagnetic materials discussed in Chapter 4. As temperature rises, thermal oscillation becomes intense and some of ε_P bonds will be broken or the bonds will vibrate more vigorously and thus the effective bond length will increase. Since this effective length is an average value, it can be stated that the effective number of ε_P bonds (for the electrical conductivity) decreases with increasing temperature.

When a voltage is applied across a conductor, the potential energy varies accordingly, and the bond orientation and length change. Let's consider the effect of thermal energy (temperature) on the electrical conductivity of solids. In Planck's law, introduced in the next section, thermal energy emits electromagnetic waves called radiation. Electromagnetic waves are electrically neutral and will not prevent the distortion of ε_P bonds by the external electric field. However, since the effective bond number decreases due to thermal vibration, the current flow rate decreases and resistance increases in the conductor. In addition, the dependence of electrical conductivity on the crystallographic orientation can be understood in terms of the nature of the interatomic PeN bond structure composed of different elements.

On the other hand, insulators maintain a stable state at

low applied voltages and relatively high temperatures because the bond strength of primary bonds is very large and there are no ε_P bonds that can carry electrical energy. If some primary bonds become thermally flexible, the electrical conductivity of insulators increases.*119) Since thermally excited ε_P bonds are isotropic, they do not induce electricity on their own, but if there is a temperature gradient in the conductor, some of the excited ε_P bonds become anisotropic to yield electrical conduction.*120)

Electrical resistance and electric wave

In general, electrical resistance of a conductor increases with temperature. Current physics explains that the electron-phonon*121) interactions are the main reasons for the resistivity. It is described by the following Bloch-Grüneisen equation at low temperatures.144

$$\rho(T) = \rho(0) + A\left(\frac{T}{\Theta_R}\right)^n \int_0^{\frac{\Theta_R}{T}} \frac{x^n}{(e^x - 1)(1 - e^{-x})} dx \quad \text{--- (5.1),}$$

---

*119) For insulators, there are few available PeNs to form ε_P bonds. Therefore, primary bonds are directly involved in the electric current. At higher temperatures, the primary bonds become enough flexible to respond to external electric fields.

*120) If one part of a metallic conductor is hot, the thermally excited bonds diffuse toward cold parts, and electricity flows due to this "thermoelectric effect".

*121) A phonon is an acoustic quantum representing the quantized vibration of a crystal lattice in solid-state physics. Phonons play an important role in the conductivity of heat and electricity of solids, and long-wave acoustic quanta produce sound waves.

where $\rho(0)$ refers to the resistivity by impurities. At very low temperatures, the total resistivity is constant, as it is mainly due to impurities. A is a constant, Θ_R is the Debye Temperature*[122] and n is an integer depending on the type of interaction. $n = 5$ for the electron-phonon scattering in typical metals, $n = 3$ for the resistivity by s-d electron scattering in transition metals and $n = 2$ is due to the electron-electron interaction. As such, it is hard to understand the behavior of electrical resistivity of a conductor depending on temperature in the regime of quantum mechanics.

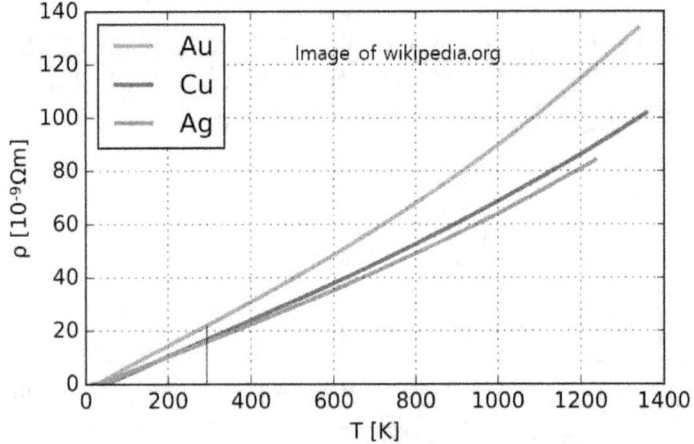

Figure 67. Temperature dependence of the electrical resistivity of gold (Au), copper (Cu) and silver (Ag).

---

*[122] It is a combination of constants that depend on the properties of the material. This is interpreted as the temperature at which the vibration mode of the maximum frequency of phonon is activated.

Table 5. Resistivity, melting point (m.p.) and modulus of sodium (Na), aluminum (Al) and copper (Cu).

	-193°C	0°C	227°C	m.p.(°C)	Modulus
Na, bcc	8.00 nΩm	43.30 nΩm		97.8	10 GPa
Al, fcc	2.45 nΩm	24.17 nΩm	49.90 nΩm	660.3	70 GPa
Cu, fcc	2.15 nΩm	15.43 nΩm	30.90 nΩm	1084.6	110-128 GPa

Consider the temperature dependence of resistivity in the regime of the new paradigm. The movement of distorted ε_P bonds in the direction of the applied voltage is hindered by thermal vibration. This hindrance is the resistivity, which increases with an increase in the frequency of collision between the distorted ε_P bonds by the electric field and the thermally excited ε_P bonds. In order for electrical energy to be transferred, interatomic links are required in the direction of the applied electric field. These links are composed of stable ε_P bonds, When they are broken or weakened by thermal vibration, the energy transfer will be more or less blocked or retarded. Therefore, the electrical resistivity will be proportional to the number of the *effective* ε_P bonds (the number of bonds having the effective length) thermally dissipated. The breakdown of the effective bonds is a thermally activated process, so at very low temperatures the resistivity will increase exponentially. Above a certain temperature, the number of effective bonds will decrease in proportion to the thermal energy, $k_B T$, so the resistivity is linearly proportional to temperature. It is thus predicted that the resistivity increases exponentially at very low temperatures and then increases linearly with temperature at some high temperatures. Figure 67 shows the temperature dependence of the resistivity for Au, Cu and Ag.[145] In general, the resistivity varies linearly with temperature, but at high temperatures it increases rapidly with increasing temperature. This means that the number of effective ε_P bonds for electrical conduction rapidly decreases by thermal vibration at high temperatures.

If electricity is substantially a wave, a medium for the propagation of wave is required. Sound waves in solids are caused by the lattice vibration. Imagine some interatomic bonds for electrical conduction, namely ε_P bonds, in a crystal lattice formed by primary interatomic bonds involving valence PeNs, electric wave can propagate along the network of ε_P bonds, which is the movement of electric field, namely the current. As the propagation speed of a wave (e.g. a sound wave) decreases with temperature due to lattice expansion, it is expected that the electrical conductivity will also decrease. The velocity v_s of a shear wave (a transverse wave) through a homogeneous three-dimensional solid is given as[146]

$$v_s = \sqrt{\frac{G}{\rho}} \quad \text{---} \quad (5.2),$$

where G is the shear modulus and ρ is the mass density. As temperature rises, the lattice expands and the density decreases, but the shear modulus decreases more rapidly, reducing its velocity as shown in Figure 68.[147] In this regard, the temperature dependence of electrical resistivity

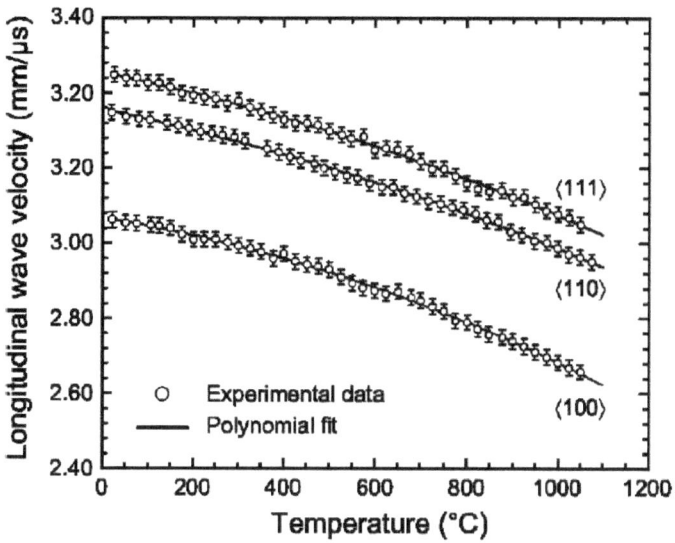

Figure 68. Temperature dependence of the longitudinal wave velocity in the <111> <100>, <110> and <111> directions of $Cd_{0.96}Zn_{0.04}Te$ (Queheillalt and Wadley).

of a conductor is expected to behave similarly to the temperature dependence of lattice constant. Figure 69 shows the thermal expansion coefficient of Cu and the temperature dependence of lattice constant (multiplied by 1,000 after subtracting the value at 20 K shown in the reference).[148] Compared with Figure 67, it is seen that the temperature dependence of electrical resistivity of Cu is very similar to that of lattice constant. In Figure 70, the thermal expansion coefficient of Au is smaller than Cu in the entire temperature range,[149] but the resistivity of Au in Figure 67 is higher than Cu. The lattice constant increased by 0.012 Å (4.067 Å to 4.078 Å) in the temperature range of 100 to 300 K for Au,[150] while it increased by 0.0106 Å from

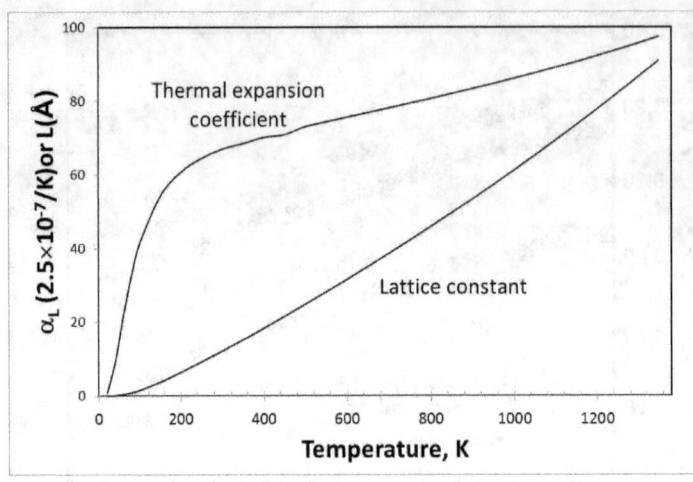

Figure 69. Temperature dependence of the lattice constant and thermal expansion coefficient of Cu (Wang and Reeber)

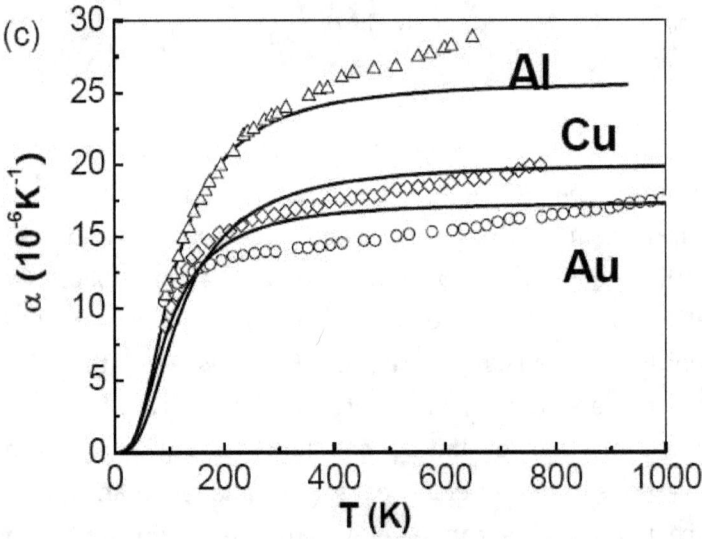

Figure 70. Temperature dependence of the thermal expansion coefficient for Al, Cu, and Au (M. Gu et. al)

3.60455 Å to 3.61515 Å for Cu.[151] Not only the absolute value of lattice constant of Cu is small, but the rate of increase is small. It can be stated that the electrical resistivity is very closely related to the lattice constant. The lattice constant increases in proportion to the increase of temperature, if Poisson's ratio defined in Figure 73 is invariant with temperature, but the lattice constant increases steeply at higher temperatures because Poisson's ratio is also a function of temperature and increases with temperature.[152] Considering the effect of magnetic field formed on the plane perpendicular to the electric field when electricity flows, the electrical resistivity as a function of temperature can also be understood from this point of view. In other words, the electrical resistivity increases linearly with temperature and steeply increases at high temperatures, similarly as the behavior of lattice constant as a function of temperature.

Electrical resistivity in the new paradigm

The electron configuration (the PeN configuration) of copper (Cu) is [Ar] $3d^{10}$ $4s^1$ and should be $3d^9$ $4s^2$ energetically. If Cu has this configuration, the internal energy is lowered. Why is the $3d^{10}$ $4s^1$ configuration? Is it due to the interaction with the solid vacuum?

The crystallographic structure of Cu is fcc.[153] If only $4s^1$ PeN is responsible for 12 bonds with the adjacent Cu atoms, the melting point and modulus of elasticity should be lower than that of Na, and the electrical resistivity should be higher. As is seen in Table 5, since Cu has a higher melting point, higher modulus of elasticity and lower electrical resistivity than Al as well as Na, it is assumed that $3d^{10}$ PeNs should be involved in the interatomic bonding. If 3 of the $3d^{10}$ PeNs are transferred to $4p^3$, a total of 4 PeNs will be involved in the bonding and thus the lattice will be stiff against thermal vibration.

For an electric current, interatomic chemical bonds have to react elastically to applied electric fields. If the distance of a bond (the distance of a valence PeN of an atom to a valence PeN of a neighboring atom) is too short, the bond flexibility will be low, and if the bond length too long, the electric wave will hardly propagate. In the case of covalent bonds, the flexibility is very low and almost fixed, so the electrical conductivity of covalently bound materials is very low.*[123] For semiconductors or insulators, the resistivity decreases with increasing temperature. In this case, lattice expansion causes the lattice to be flexible and if the flexible lattice responds to the electric field, the electrical conductivity will increase.

However, in order for electricity to flow in the absence of free electrons, the formation and movement of electric fields through the deformation of ε_P bonds are essential. When electricity flows, a magnetic field is induced perpendicular to the direction of current. If an electric field is formed via distortion of ε_P bonds, the induced magnetic field is another distortion in the plane perpendicular to the electric field. The electrical conductivity will be high, if the rearrangement of PeNs in ε_P bonds occurs in the direction

---

*[123] While diamond has a very high electrical resistivity, graphite has a high electrical conductivity compared to diamond because there is a valence PeN per atom that does not participate in the primary stiff covalent bond.

in which the magnetic field in the vertical plane well develops (elastically). Another condition for a good conductor is that this electric field must move well. That is, the distorted ε_P bonding structure must move freely and fast. Water waves move on the water surface because all the water molecules are connected by hydrogen bonding. In an external electric field, the PeNs are relocated and the associated ε_P bonds are accordingly rearranged, but still thermally oscillate constantly. The transmission of this biased vibration in one direction is the current.*[124]

Finally, let us check for the two copper oxides if the mechanical and electrical properties vary depending on whether valence PeNs are involved in primary bonds or secondary bonds required for electrical conduction, ε_P bonds.*[125] Figure 71 shows the crystal structure of two stable copper oxides. Cu_2O (specific gravity 6) has four Cu atoms bound to one oxygen (O) atom in a diamond form, and each Cu atom combines with two O atoms. CuO (specific gravity 6.315) has four Cu atoms bound to one O atom, but each Cu atom again is bound to four O atoms. The specific gravity of CuO having a large number of primary bonds should be larger than that of Cu_2O.

If primary bonds determine the melting point, then the melting point must be high for CuO, and Cu_2O with two

---

*[124] Heat energy can be converted into electrical energy in solids. This is a thermoelectric effect. If one side of a solid is hot and the other side is cold, the crystal structure of the solid will be anisotropically distorted along the thermal gradient and the movement of thermal vibration will also have a anisotropy.

*[125] Copper oxide is a very important one in high temperature superconductivity. See Chapter 6.

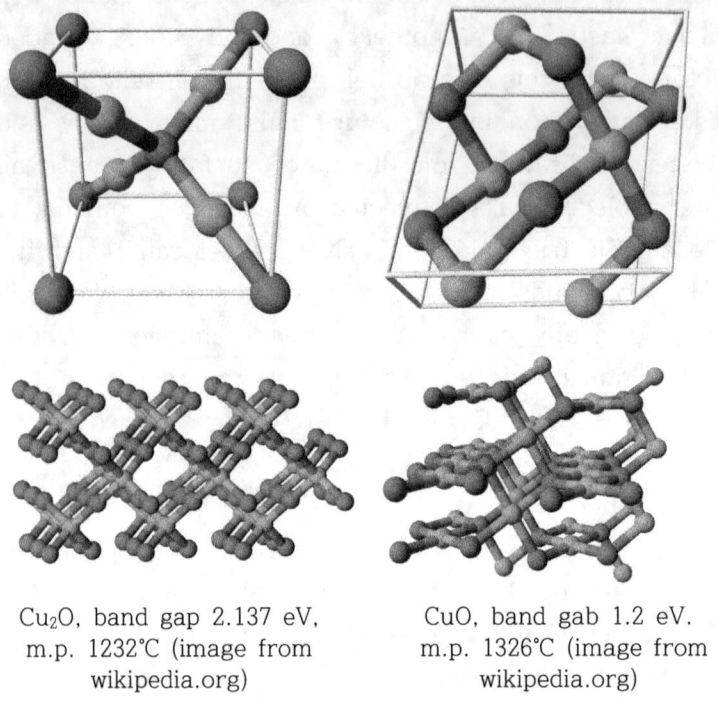

Cu₂O, band gap 2.137 eV, m.p. 1232°C (image from wikipedia.org)

CuO, band gab 1.2 eV. m.p. 1326°C (image from wikipedia.org)

Figure 71. The crystal structure of two kinds of copper oxides

extra PeNs must have a higher band gap in terms of ε_P bonds. In fact, these predictions are correct. In conclusion, the crystal structure and the electromagnetic characteristics vary depending on whether the valence PeNs participate in primary bonds or secondary bonds. The superconducting phenomena discussed in Chapter 6 well develop in unstable crystal structures in which these primary and secondary bonds are not clearly defined. This unstable structure is very flexible and elastic regarding to ε_P bonds but solidifies into one phase below the critical temperature.

5.3. Magnetism in the new paradigm

The elementary unit of magnetic field is electron spin. What is electron spin and does it really exist? In quantum mechanics and particle physics, spin is an internal form of angular momentum of elementary particles, composite particles (hadrons), and atomic nuclei.[154] In quantum mechanics, there are two kinds of angular momentum: spin and orbital angular momentum. Orbital angular momentum occurs when a particle orbits a trajectory like an electron orbiting an atomic nucleus.[155] Spin angular momentum can be found, for example, in the Stern-Gerlach experiment, in which particles with angular momentum that are not explained by orbital angular momentum alone were observed.[156]

Stern-Gerlach experiment

This experiment was performed by Stern*[126] and Gerlach*[127] in 1922 and confirmed that the spatial orientation of angular momentum is quantized. They measured the trajectory of silver atoms evaporated in a furnace after passing through a magnetic field, as shown in Figure 72. If a particle has a magnetic moment related to spin, it will be affected by the magnetic field. Experimental results showed that the silver atoms settled down to specific final positions, depending on the magnetic field strength. In 1927, an

---

*126) Otto Stern (2.1888 - 8.1969) was a German physicist and worked in America since 1933. He won the Nobel Prize in Physics for the Stern-Gerlach experiment in 1943.
*127) Walther Gerlach (8.1888 - 8.1979) was a German physicist.

experiment on hydrogen atoms showed the same results as in Figure 72.[157] This is due to the quantum mechanical properties of spin, and thus it was demonstrated that spin angular momentum is quantized at the atomic level.[158]

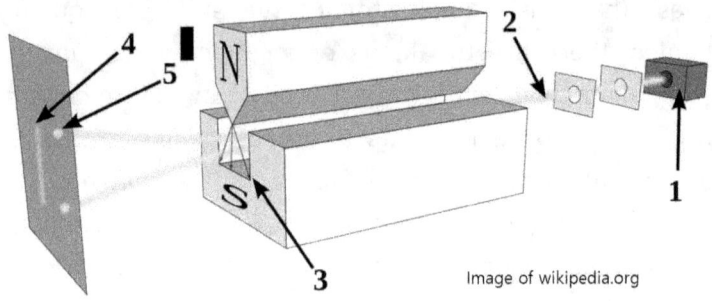

Image of wikipedia.org

Figure 72. Stern-Gerlach Experiment: When silver atoms pass through a non-uniform magnetic field, they bend upward or downward due to the spin effect: 1. electric furnace; 2: silver atom beam; 3: magnetic field. 4: expected trajectory; 5: measured trajectory.

Spin is a type of vector that has a magnitude and quantized orientation. All elementary particles of one kind have the same spin angular momentum.[159] In conjunction with the spin-statistics theorem, electron spin gave rise to the Pauli exclusion principle, which in turn became the basis of the periodic table of elements. Pauli first introduced the concept of spin, theorized mathematically in 1927 and included it in his relativistic quantum mechanics in 1928.[160]

Electromagnetic field in the new paradigm

Previously, electricity and magnetism were interpreted as separate physical phenomena. Maxwell's electromagnetic theory and its verification made them inseparable.[161] The theory of electromagnetism applies perfectly to optics and electrical circuits. The change in the electric field cause changes in the magnetic field, and the change in the magnetic field yields changes in the electric field. To know the magnetic field, we have to consider the movement of electrons (or the movement of electrical energy carried by electrons). In the new paradigm, however, there are no electrons as independent particles in solids.*[128] Free electrons in a conductor are replaced by the distorted and excited secondary bonds and the movement of electrons is nothing but the movement of this excited bond structure, that is, the transportation of energy stored in the distorted ε_P bonds in a wave form. Since there are no free electrons in the new paradigm, the movement of electrons in a conductor is interpreted as a wave in which the interatomic (bond) energy changes periodically. The interatomic bonds are ε_P bonds involving two valence PeNs of two adjacent atoms and the flow of electricity is nothing but the flow of distortion energy of ε_P bonds.

So why do the atoms in a conductor form ε_P bonds? As can be seen from the above Stern-Gerlach experiment, there is also quantized angular momentum at the atomic level in the interiors of a conductor, and the whole atomic angular momentum (the whole atomic magnetic field) of the

---

*[128] As already mentioned, a free electron is a special form of electromagnetic wave, as shown in Figure 3.

conductor will affect the behaviors of individual valence PeNs to maintain a certain order. This is the origin of ε_P bonds, which are repeatedly disintegrated and recombined by thermal vibration. As in the Stern-Gerlach experiment, if there is an external field, the valence PeNs will follow it and move. ε_P bonds are rearranged along the external electric field so that random vibrations become anisotropic and additional vibrations in the plane perpendicular to this direction will spread out. This is the magnetic field.

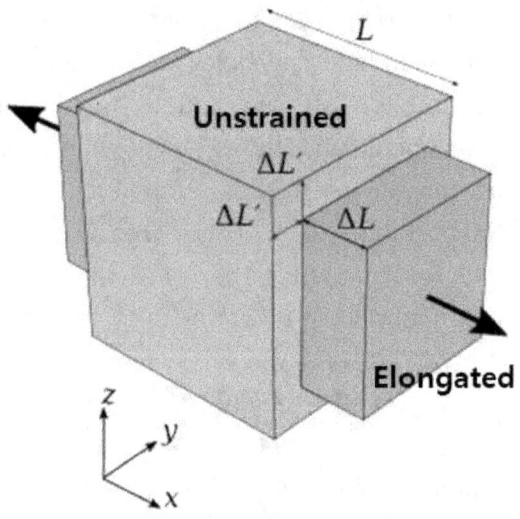

Figure 73. When a material is tensioned (or compressed) in one direction, it is compressed (or elongated) in the plane perpendicular to that direction. Poisson's ratio is the ratio of the compression to the elongation. When the length of one side of the cube increases in the x direction via elastic deformation, the length of the side decreases in the perpendicular directions (y, z). For a small deformation Poisson's ratio is given as $-\Delta L'/\Delta L$ (from wikipedia.org).

The magnetic field always forms in the plane perpendicular to the electric field. It is similar to the mechanical behavior of solids in which a unidirectional stress acting on tension (or compression) induces oppositely compressive (or tensile) stress in the perpendicular directions. The ratio of the contracted length to the elongated length is called Poisson's*[129)] ratio (Refer to Figure 73). While Poisson's ratio is the reaction of stress acting on primary interatomic bonds, the generation of a magnetic field by an electric field is the reaction due to the deformation of ε_P bonds.

The electron probability density of the hydrogen atom is very similar to the stress field around a dislocation in a metallic crystal (Figures 25 and 26). The energy level of electron represents the magnitude and state of stress exerting on the solid vacuum formed by the atomic mass of hydrogen. Thus, the movement of electrons in a conductor means a periodic change in this stress field strength. If the strength changes at one point, the nearby stress field is influenced. If the strength changes anisotropically, the surrounding stress field will change accordingly. Since the electron flow is an energy flow and the energy is stored in the stress field, the magnetic field can also be interpreted as a change in the surrounding stress field. The free electrons in a conductor are replaced by ε_P bonds so that the periodic stress change due to the distortion of the bonds on the atomic level is the current.

---

*[129)] Siméon Denis Poisson (6.1781 – 4.1840) was a French mathematician, engineer and physicist. He was the last to oppose the wave characteristics of light at that time.

5.4. Maxwell equations in the new paradigm

Maxwell's equations are partial differential equations that, together with Lorentz force law,*[130] form the basis of classical electromagnetism, optics, and electrical circuitry. They describe the generation of electric and magnetic fields by electric currents and charges, and prove that the waves of electric and magnetic fields propagate at the speed of light in the vacuum. They were first published by Maxwell around 1861 and predicted that light is an electromagnetic wave. The electromagnetic behaviors of matter are described in relation to the total electric charge and total current of matter. Although introduced in Section 4.1, they are introduced once more with the interpretations in the regime of the new paradigm.

New interpretation of Maxwell's equations

These equation are based on the four laws of electromagnetism.
- Gauss' law: A current leaving a space is equal to the amount of charge.

Interpretation: A current is the transportation of the distortion energy of ε_P bonds. A current out of a space means that distorted ε_P bonds propagate to the outside. A charge quantum is a distorted ε_P bond and charge is the

---

*[130] It describes the force a charged object feels in electric and magnetic fields. Positive particles accelerate by the electric field and move helically as they pass through the magnetic field. This law was derived by Dutch physicist H.A. Lorentz (7.1853 - 2.1928).

potential energy of this distorted bond structure. The distortion energy of ε_P bonds can be expressed in the form of charge.

- Gauss's law for magnetism: There is no magnetic monopole. The net magnetic field flowing out of a closed surface is zero.

Interpretation: A magnetic field is an induced electromagnetic stress field formed only by an electric current (the flow of electric field). Thus, magnetic nonopoles cannot exist. Because it is generated in response to the electric field, the magnetic field entering and exiting a closed surface (a finite space) must be zero. It is not created or destroyed by itself. This is because the internal electric field is due to the distortion of ε_P bonds and the magnetic field is an induced distortion of ε_P bonds.

- Faraday's law of induction: The voltage induced in a closed loop is proportional to the rate of change of magnetic field strength surrounding the closed loop.

Interpretation: The change in the magnetic field strength means that the number of distorted ε_P bonds increases or decreases in the perpendicular directions to the flow of electricity. Increasing the number of distorted states caused by the magnetic field means that the number of distorted ε_P bonds in the conductor of the closed circuit increases. If this condition is not eliminated immediately, it appears in the form of voltage.

- Ampére's circuital law: The magnetic field induced around a closed circuit is proportional to the current and the rate of change in the electric field (enclosing the closed circuit).

Interpretation: This is the symmetrical law of Faraday's

electromagnetic induction law. A current is the movement of electric field. An electric field is formed due to the distortion of ε_P bonds, and the movement of this distortion induces a secondary distortion called the magnetic field in the perpendicular plane. Flow of electricity means that the distorted bonding state moves in one direction and is represented by the intensity of the induced distortion called the magnetic field.

Maxwell's equations match well with Einstein's theory of relativity. Actually Einstein developed the theory of relativity that only relative motions yield physical results using the absolute speed of light resulting from the Maxwell's equations.[162] The equations explain a wide variety of macroscopic physics, including the theory of relativity, but they are not perfect and in some cases very inaccurate. They do not hold in very strong electromagnetic fields and at very short distances (due to vacuum polarization). Phenomena such as nonclassical light*[131] or quantum entanglement of electromagnetic fields are impossible to interpret in the Maxwellian regime. Phenomena involving individual photons, such as photoelectric effects and Planck's law,*[132] are difficult or impossible to explain.

---

*131) This cannot be explained by Maxwell's electromagnetism, but by quantum electromagnetic field theory and quantum mechanics. A typical nonclassical light is squeezed light. Its change in amplitude and wavelength cannot be classically described.

*132) This is a law that describes the spectral density of electromagnetic waves emitted by a black body at a given temperature. It was proposed by Max Planck (4.1858 - 10.1947) in 1900 and contributed greatly to the birth of quantum mechanics. Planck won the Nobel Prize in Physics in 1918 for that.

These phenomena are addressed by quantum electrodynamics. Maxwell's equations show inconsistencies in photon-related phenomena such as photon-photon scattering, quantum optics, and reveal limitations of classical theory along with the Newtonian mechanics.

In physics, a magnetic field is generated by the movement of electrons. Maxwell's theory would not hold if electrons were created or destroyed. According to the new paradigm, a free electron is a kind of distortion of the solid vacuum as a circular standing electromagnetic wave. If this distortion moves, the deformation field of the nearby solid vacuum also moves. An electron in rest also generates a magnetic field by spin originated from the circular movement of the standing wave. In terms of distortion of the solid vacuum, extending Maxwell's equations may also explain how electrons are created or destroyed. This book introduces some of the quantum mechanical phenomena that Maxwell's equations cannot explain and interprets them in relation to the properties of the solid vacuum in the new paradigm.

Quantum entanglement and vacuum polarization

Quantum entanglement is a phenomenon that in any quantum mechanical systems, individual quantum states are entangled with each other and when one quantum state changes, another entangled quantum state changes instantly (at or above the speed of light). This phenomenon is reproduced in entangled systems even when the physical distance between the two quantum states in the target

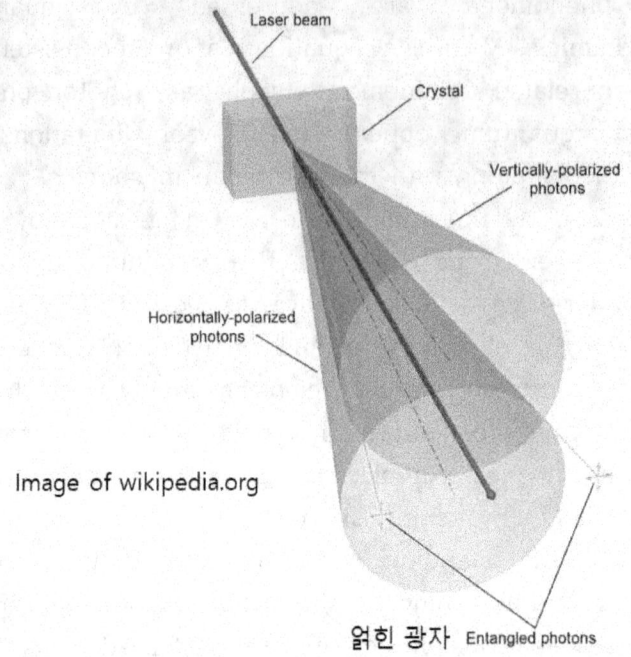

Figure 74. Spontaneous parametric down-conversion process produces a pair of entangled photons polarized perpendicularly to each other.

system is very large. The distance may be as large as that from the Brazilian capital city to Seoul, Korea. Quantum entanglement phenomena have been observed in electron pairs as well as photons.[163] The spin state of one electron at rest will be delivered at the infinite speed, and immediately transferred to the other entangled electron.*[133]

Recently, China announced that it has succeeded in long distance (1,200 Km) data transmitting using the principle of

---

*[133] Quantum entanglement will be handled in detail in the book "Particles" published in near future.

quantum entanglement.*134) 8 million pairs of photons allegedly participated in this transmission, and the speed was 1 trillion times faster than that by conventional optical cables. Why are these entangled pairs created? How are these photon pairs linked together? This suggests that the vacuum is a space full of something. The infinite propagation speed means that the way of information transmission is not through the transmission in a conventional wave form and reminds us of matter wave. The speed of a matter wave is infinite when matter is at rest, as the author also mentions in "Origin of Gravity and New Cosmos."

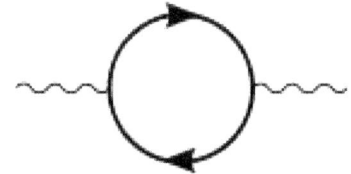

Figure 75. Concept of vacuum polarization (image from wikipedia.org)

While quantum entanglement is to show the change of quantum states over long distance, vacuum polarization is about the polarization of the vacuum at very short distance. In quantum field theory and quantum electrodynamics, vacuum polarization means that an electromagnetic field produces a virtual electron-positron

---

*134) "Progress in Commercializing Quantum Communications⋯ China, World's first quantum entanglement transmission success", 연합뉴스 (Yonhap News Agency) 2017/06/17 11:03.

pair, altering the distribution of charge and current. It is also called the self-energy of force carriers (gauge boson, photons, etc.).*135) In quantum theory, a short-lived "virtual" particle-antiparticle pair between two interacting particles in the vacuum is created with energy corresponding to the time following Heisenberg's uncertainty principle*136) and disappears after that time. This particle-antiparticle pair can be quarks, gluons, electrons, and the like. A hypothetical electron-positron pair acts as an electrode and rearranges in the electromagnetic field around the electron. This rearrangement causes the electromagnetic field applied to the outside to be weaker than when there is nothing in the vacuum. This is vacuum polarization.164 The lifetime and energy of the electron-positron pair is determined locally from the relation of time and energy according to the Heisenberg uncertainty principle.

Maxwell's theory may also be applied to quantum entanglement for a system that contains lines connected to

---

*135) In particle physics, gauge bosons are force-carrying particles with a spin of one. Elementary particles, described by gauge theory, interact through the exchange of gauge bosons called virtual particles. Virtual particles are instantaneous quantum fluctuations based on the principle of uncertainty and have some characteristics of ordinary particles. In quantum field theory, the interaction between ordinary particles is explained by the exchange of virtual particles. Ordinary particles and virtual particles need not have the same masses, but the longer the lifetime of virtual particles is, the closer they are to the characteristics of normal particles. In quantum field theory, the attractive and repulsive force between two charged particles is considered to be due to the virtual photon exchange. Virtual photons are particles that are exchanged in electromagnetic interactions.

*136) German physicist W.K. Heisenberg discovered this principle. Refer to foot note *5).

two quanta, which is like a solid bar (similar to the characteristics of the solid vacuum). The change on one side leads directly to the change on the other. Vacuum polarization is a phenomenon originated from the uncertainty principle. It exists only for a very short time, and has nothing to do with macroscopic Maxwell's theory. However, if the solid vacuum has a regular lattice structure as shown in Figure 1, vacuum polarization would be observed if very short wavelengths of light are absorbed in the vacuum structure. This is a kind of diffraction of light by the solid vacuum which allows to reveal the solid vacuum structure. Vacuum polarization can be imagined to be originated from the structure of the solid vacuum and that the vacuum is not just an empty space but made of a very dense medium. As an electrical current is the flow of distorted ε_P bonds, vacuum polarization may be regarded as a phenomenon due to the distortion of the solid vacuum structure, which may be approached consistently via Maxwell's equations.

Photoelectric effect

The photoelectric effect is a phenomenon that Maxwell's laws do not hold. When light is irradiated to a material, electrons or other charged particles are emitted as shown in Figure 76. The emitted electrons are called photoelectrons. This phenomenon was discovered at the end of the nineteenth century. In order to explain this effect, Einstein published a paper in 1905 stating that light energy is quantized. In 1914 Millikan*[137]) validated the Einstein model with an oil drop experiment*[138]) using a device like that shown in Figure 77.

Although research on the interaction between light and the solid vacuum is not easy to find, we can assume that there are quite a lot of works going on in this filed since a paper called "Nonlinear Interaction of Laser Light and Vacuum" was presented at a conference in 2016.[165]

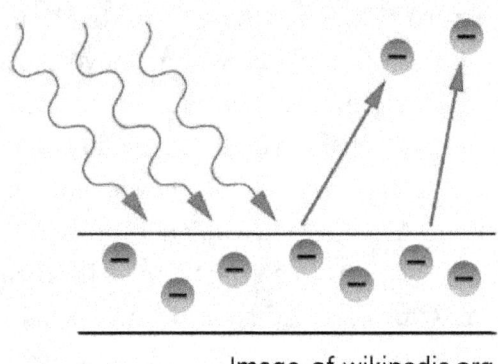

Image of wikipedia.org

Figure 76. Electrons are emitted by the photoelectric effect.

---

*137) R.A. Millikan (3.1868 – 12.1953) was an American experimental physicist. In 1923 he won the Nobel Prize in Physics for his work on the elementary electric charge and photoelectric effect.

*138) When oil mist flowing through small holes in the top plate into the space separated by two metal plates, the mist particles can be ionized and negatively charged by x-rays. Applying a voltage between the two metal plates to attract the oil droplets to the top plate will find the voltage that is in equilibrium with the gravitational force. Through this, it is possible to estimate the charge of oil droplets, and the measured charge value was the multiple of 1.5924×10^{-19}C. This value is only 0.6% different from $1.6021766208 \times 10^{-19}$ C, the elementary charge of electron.

According to classical electromagnetic theory, the photoelectric effect is a phenomenon that occurs because light energy is transferred to electrons in the material. Varying the light intensity can change the kinetic energy of photoelectrons. So no matter how dim the light beam is, the electrons can accumulate energy to escape from the material. This was different from the actual photoelectric effect. Instead, electrons were emitted only for light with above a critical frequency, and below that, no electrons were emitted regardless of the light intensity or irradiation time. Einstein stated that this is because light is not just a wave but is composed of discrete quanta (photons), which supported Planck's energy quanta previously discovered.[166]

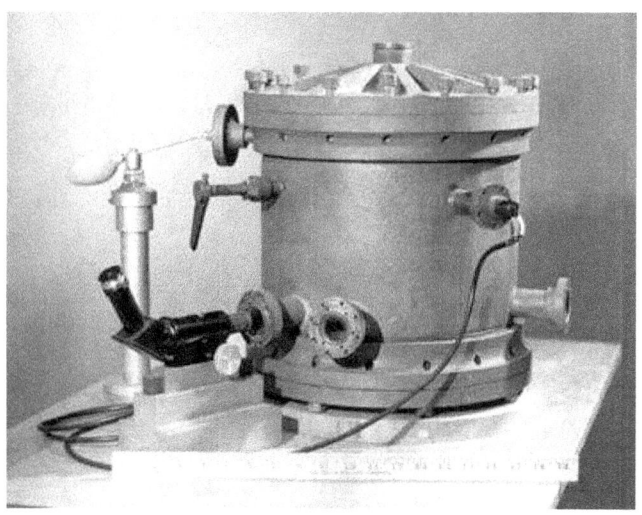

Figure 77. Millikan's Oil Drop Tester of 1914.

Photons have specific energy proportional to the frequency of light. When irradiated, the electrons in the material absorb the energy of photons and are released when the energy exceeds the work function (the electron bonding energy) of the material. The emitted electron energy is independent of the light intensity and depends only on the individual photon energy, or frequency. In quantum mechanics, this is due to the interaction of light with valence electrons. All the photon energy must be absorbed and used to emit an electron, otherwise the energy is reflected. If the photon energy is greater than the energy used to release the electron, the rest energy becomes the electron kinetic energy.[167] There is a minimum frequency of light that causes the photoelectric effect. Fixing the number of photons and raising the frequency increases the maximum kinetic energy of photoelectron as shown in Figure 78.

Increasing the incident beam intensity increases the amount of resulting current, because the number of emitted electrons increases. The time it takes for an electron to emerge after the irradiation is less than 10^{-9} seconds. The direction of photoelectrons depends on the polarization of the electromagnetic waves (the direction of the electric field when linearly polarized).[168] The maximum kinetic energy of photoelectron is

$$K_{max} = hf - \phi \quad \text{---} \quad (5.3),$$

where h is Planck's constant and f is the frequency of the incident light. φ is called the work function and is the minimum energy to yield a photoelectron from the metal

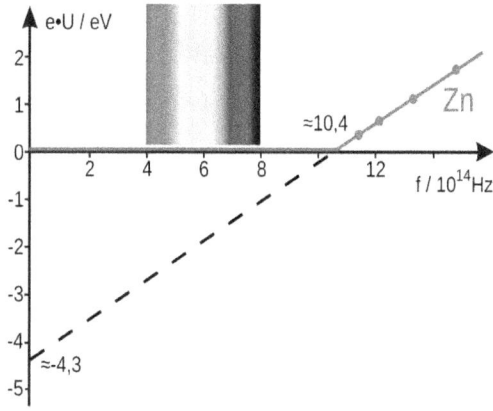

Figure 78. Maximum kinetic energy of photoelectron as a function of the frequency of light irradiated on zinc (image from wikipedia.org).

surface. In terms of the critical frequency f_0

$$\phi = h f_0 \quad \text{---} \quad (5.4).$$

Therefore

$$K_{\max} = h(f - f_0) \quad \text{---} \quad (5.5).$$

Using a device like that shown in Figure 79, the maximum kinetic energy of photoelectron can be determined. When a light beam from the lamp passes through the filter and hits the curved electrode, electrons are emitted. Applying a negative voltage to this electrode reduces the current caused by the photoelectric effect and becomes zero at a certain value. This voltage is called the stopping voltage V_0.[169] With q_e the electron charge, the energy that traps a photoelectron is then $q_e V_0$, which is the maximum kinetic energy K_{max} of photoelectron when the applied voltage is zero. In other words,

$$K_{max} = q_e V_0 \quad \text{---} \quad (5.6).$$

From Eqs. (5.5) and (5.6), it is seen that the stopping voltage varies linearly with the frequency of light and depends on the type of material, that is, the work function.

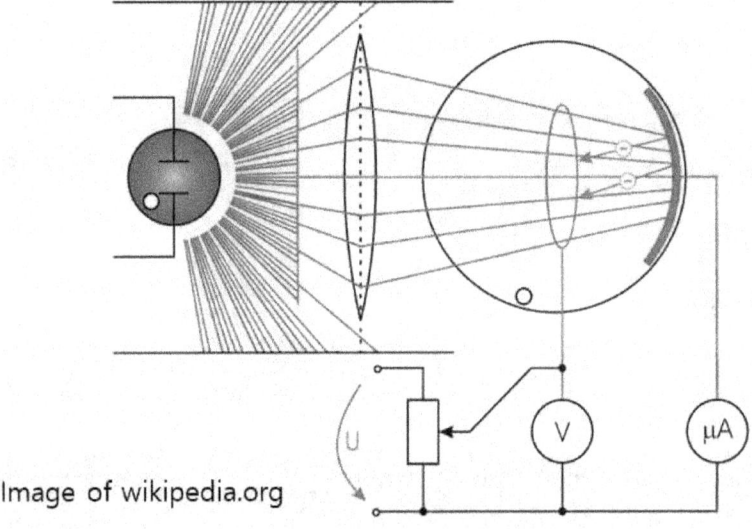

Image of wikipedia.org

Figure 79. Schematic of the experimental apparatus for verifying the photoelectric effect. When a light beam from the lamp passes through the filter and hits the curved electrode, electrons are released. When a voltage is variably applied to the electrode, the voltage at which electricity does not flow, the stopping voltage, is obtained.

The photoelectric effect was a very important discovery that led to the birth of quantum mechanics, explained by the introduction of light quanta, or photons. A fundamental premise that explains this effect is that light is irradiated on a solid, causing the electrons "bound" to the atoms in solids. Photoelectrons are observed or measured externally in free space but it is not known whether they belong to the constituent atoms or not. If the total number of electrons in the system is preserved, then there will be a lack of electrons in the solid as much as the number of electrons released. But what if electrons were made from light (rather than the light energy being transferred to the solid, but from the solid)? Since light is electrically neutral, positrons have to form when electrons are made. If the electron in the solid is a deformed ε_P bond, the positron is also a deformed secondary bond that is antisymmetric to this bond. The total charge must be zero, whether the irradiation of light emits electrons from the solid or the electrons are made from the light waves being irradiated. The hypothesis that electrons are produced from incident light waves on the solid surface rather than being emitted from the solid is consistent with our new atomic model and the concept of electron-free current, and is also consistent with the concept of Figure 3, where an electron is a circular standing light wave. There is no distinction between the electrons emitted from the solid and the electrons produced from incident light beams on the solid surface. It can be stated that the conditions for an incident straight light wave to become a circular standing wave as shown in Fig. 3 are satisfied on the surface. We imagine if the frequency of light waves and the vibration of a ε_P bond in the solid cause resonance, the light energy is absorbed by the bond and the bond energy is released again in the form of electron. This is of course a different concept than conventional quantum mechanics, but some recent studies have reported the formation of electron-positron pairs from laser beams (Laser-electron collision,[170] Photon-photon collision).[171] The photoelectric effect supports the idea that light can be transformed to free electrons by interacting with ε_P bonds at the surface of a solid.

Plank's law

Plank's law is a law on the spectral density of electromagnetic radiation from a blackbody that is in thermal equilibrium at a given temperature, and there should be no exchange of matter with its surroundings.[172]

At the end of the nineteenth century, the behavior of the spectra from black bodies at high frequencies (short wavelengths) could not be theoretically explained at that time. As shown in Figure 80, the radiant energy density at the short wavelength ranges predicted by classical theory is very different from the observed. This discrepancy was known as the ultraviolet catastrophe. In 1900 Max Planck derived a rule of thumb for the blackbody spectrum, stating that the energy of charged oscillators in a blackbody does not vary continuously but discontinuously in some minimum multiples. This minimum amount is proportional to the energy of electromagnetic waves. This law solved the ultraviolet catastrophe of the time and was the pioneering insight that underlies quantum theory.

All physical bodies spontaneously radiate heat whose

spectral radiance is expressed in terms of their area, emission angle and energy density per frequency. The energy density B_ν of the spectrum emitted at temperature T and frequency ν is given as

$$B_\nu(\nu, T) = \frac{2h\nu^3}{c^2}\left(e^{\frac{h\nu}{k_B T}} - 1\right)^{-1} \quad \text{--- (5.7)},$$

where k_B is Boltzmann's constant, h is Planck's constant, and c is the speed of light in the body. In terms of the wavelength λ, the energy density B_λ is given as

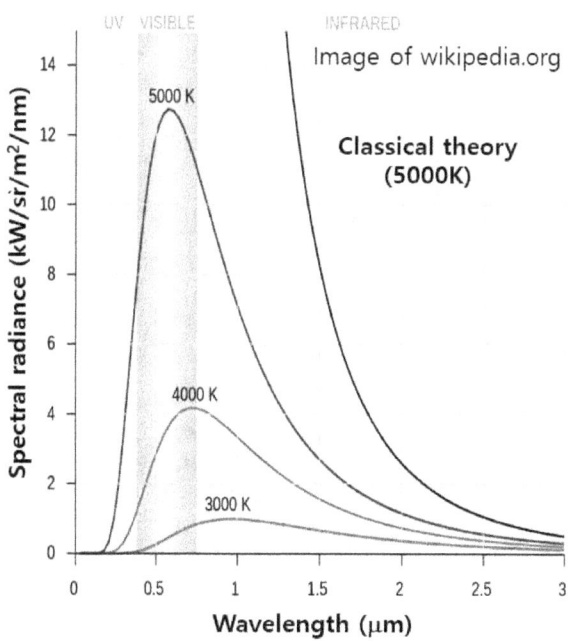

Figure 80. Blackbody radiation energy density according to Planck's law

$$B_\lambda(\lambda, T) = \frac{2hc^2}{\lambda^5}\left(e^{\frac{hc}{\lambda k_B T}} - 1\right)^{-1} \quad \text{---} \quad (5.8).$$

Accordingly, the higher the temperature, the shorter the wavelength and the faster the energy release.[173]

Black bodies are ideal objects that absorb or emit radiation of all wavelengths. Radiation heat from a blackbody in thermodynamic equilibrium is explained by Planck's law, and the higher the temperature of a body, the more radiation it emits at all wavelengths.

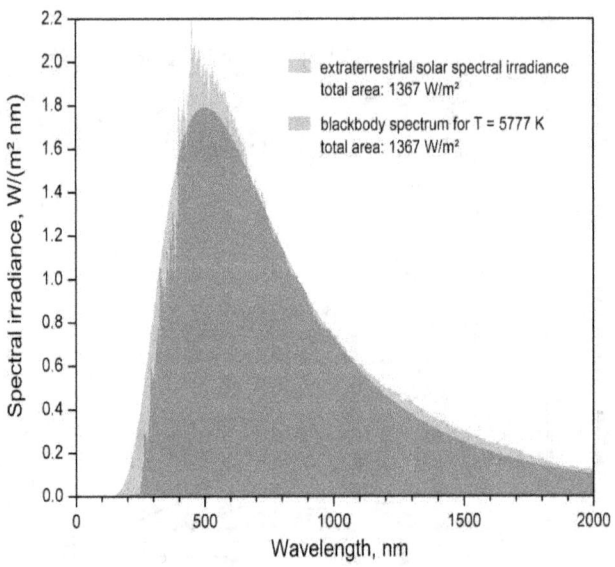

Figure 81. Energy spectrum from the Sun. It is similar to the blackbody radiation at 5,777 K (image from wikipedia.org).

The wavelength at which the maximum radiant heat is emitted depends on the temperature of the body, as shown in Figure 80. Radiation at room temperature (~ 300 K) is mostly invisible because the wavelength is in the infrared region. When temperature rises, it can be perceived as heat and red light. At high temperatures, the wavelength shortens and the body emits ultraviolet and x-rays, and appears to be bright yellow or light blue, The solar surface emits large amounts of ultraviolet and infrared rays as shown in Figure 81 and the maximum radiation is in the visible light range.

Planck's radiation is the maximum radiant heat from the surface of an object in thermodynamic equilibrium and is not relevant to the chemical composition or surface structure.[174] The amount of radiant heat passing through the interface of a medium depends on the emissivity (the ratio of actual radiation to Planck's radiation) ε at that interface and is a function of the chemical composition, physical structure, temperature, wavelength, angle of radiation, and polarization.[175] ε has a value between 0 and 1. An ideal blackbody has an ε of 1 and absorbs all radiations. The surface of a blackbody can be thought of as a hole of a large container surrounded by opaque walls and kept at a constant temperature. Radiation within this container follows Planck's law, and radiation exiting this hole follows the same law. While the Maxwell-Boltzmann distribution*139) is the maximum entropy and energy

---

*139) A probability distribution. It was first introduced to show the velocity distribution of particles (atoms or molecules) in an ideal gas in thermodynamic equilibrium. The statistical distribution of velocity in this distribution can be obtained from the equation of particle energy and kinetic energy.

distribution of particles in thermal equilibrium, the Planck distribution is for photons.[176] For gas particles, the number and mass are important variables, but for radiation, the pressure and energy density of photons in thermal equilibrium are only a function of temperature. If the distribution of photons deviates from Planck's law, photons are generated or destroyed by interaction with photons and other particles according to the second law of thermodynamics to achieve thermodynamic equilibrium. The distribution at this time is the Planck distribution.

Planck's law cannot be handled in terms of classical mechanics because it causes the ultraviolet problem that the total radiation of a black body is infinite according to the equipartition theorem.*[140] Particles with masses can be handled in quantum mechanics, but for massless ones, such as photons, quantum mechanics is not enough and replaced by quantum field theory.

In quantum field theory, photons are regarded as mass- and charge-less bosons and mediate electromagnetic forces between charged particles. The number of photons is not preserved. Photons are generated and destroyed to fill the blackbody according to the Planck distribution. When

---

*140) This theorem deals with the relationship between the temperature and average energy of a physical system. When the system is in thermal equilibrium, the energy is distributed evenly in any form. For example, the average kinetic energy per degree of freedom in translational motion must be the same as that of rotational motion. Using this theorem, the average of the total potential energy and the total kinetic energy at a temperature can be obtained to give the heat capacity of the system. It can be applied to any complex system in thermodynamic equilibrium and can predict the ideal gas law or stellar properties.

photons are in thermodynamic equilibrium, the internal energy density is completely determined by temperature. And the pressure is determined by the internal energy density. This is different from the normal thermodynamic equilibrium, in which the internal energy depends not only on temperature, but also on the number of different molecules and their molecular properties. Individual molecules have their own energy systems (the specific heat).

In the new paradigm, the electrons in a conductor are just the vibration of distorted ε_P bonds and the vibration of undistorted neutral ε_P bonds is a thermal vibration. Undistorted bonds can be distorted when an external mechanical stress is applied, so that electricity can flow. This effect is called the piezoelectric effect. The mechanical energy is stored in ε_P bonds and flows in the form of electricity. When an external electric field is applied, the thermal random vibration becomes anisotropic and electricity flows because there develops a gradient in the density of anisotropically distorted ε_P bonds along the external electric field. In the absence of external electric fields, the radiation follows Planck's law, but if an external electric field is applied, the radiation will be distorted and emitted in the form of thermal electrons. The electron is substantially a special form of light, and if we include this concept in the Maxwell's equations, we may overcome their limitations.

VI. New paradigm for superconductivity

Superconductors, the dream materials in which electricity flows without resistance, show their superconductivity at very low temperatures. Temperature should be lowered at least to the boiling point of liquid nitrogen. If superconductors that are active at room temperature are found or synthesized and applied to real life, the heat problems caused by electrical resistance are naturally solved, and a great impact on our lives is expected. In order to obtain room temperature superconductors, it is very important to know the fundamental mechanism of superconductivity. Unfortunately, modern physics has not yet given any reliable answers. The theory of low temperature superconductivity in metals appears to have been successful, but it is not sufficient to explain high temperature superconductivity in oxide based superconductors. Based on the new definition of electromagnetism in the previous chapters, we interpret superconductivity in the regime of new paradigm that an electric current is not the flow of electrons but the propagation of interatomic bond structures upon responding to electromagnetic fields. Thereon, we discuss the possibility of room temperature superconductors, the dream materials.

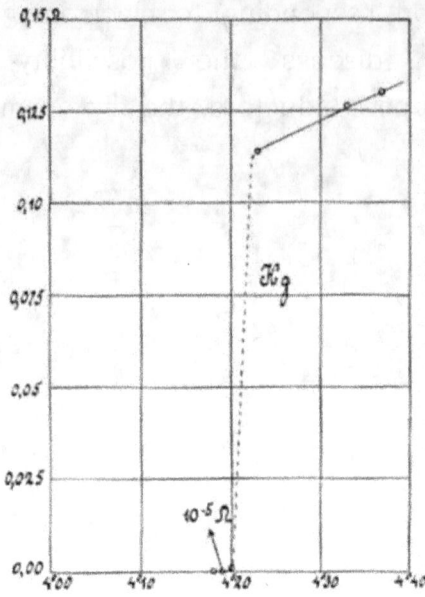

Figure 82. Superconductivity: Zero resistivity below the critical temperature (first discovered by Onnes in 1911).

6.1. Superconductivity and the BCS theory

Superconductivity is a phenomenon in which the resistivity of a conductor (a semiconductor or an insulator for high temperature superconductors, HTSs) becomes zero at low enough temperatures (see Figure 82) and the internal magnetic field is expelled (see Figure 83). Solids composed of metallic elements or alloys exhibit superconductivity at very low temperatures (critical temperatures, T_c) below 30 K. The phenomenon was first discovered for mercury by

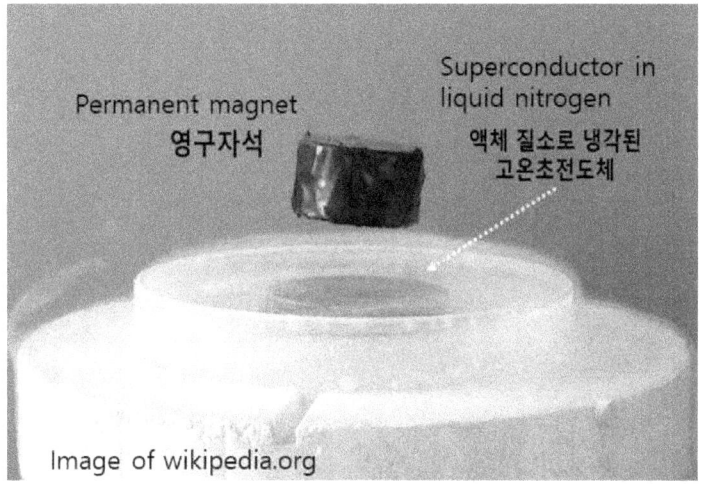

Figure 83. Superconductors repel internal magnetic fields. Phenomenologically, the superconductor pushes the permanent magnet to levitate against the gravitational force.

Dutch physicist Onnes*141) in 1911.[177] Since then, superconductors with high T_c have been found, as shown in Figure 84,[178] and the current critical temperatures reach close to 200 K.

According to the new paradigm, the flow of electricity is not the movement of particles called electrons, but the movement of distorted secondary bonds, namely ε_P bonds.*142) These distorted bonds have potential energy, since the valence PeNs of ε_P bonds are deviated from their

---

*141) Heike Kamerlingh Onnes (9.1853 - 2.1926) was a Dutch physicist. He liquefied helium for the first time in 1908 and used it to measure the resistivity of mercury and discovered superconductivity in 1911. For this, he won the Nobel Prize in Physics in 1913.

*142) A ε_P bond is a chemical bond between a valence PeN belonging to one atom and a PeN of a neighboring atom.

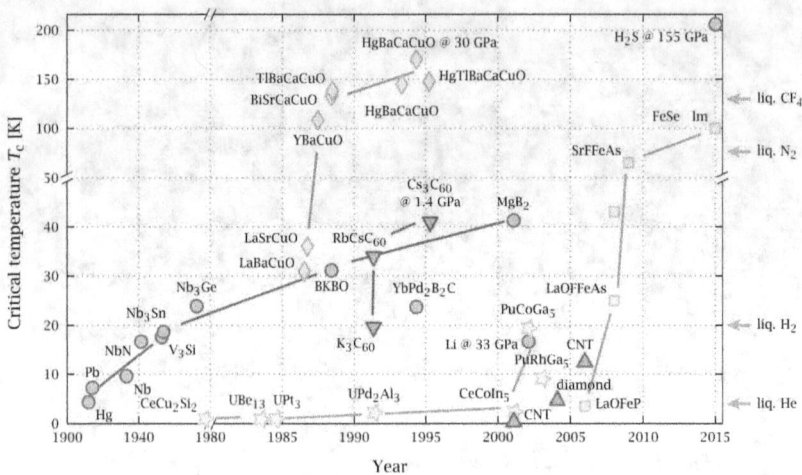

Figure 84. Development of superconductors (P.J. Ray)

thermal equilibrium positions due to the presence of electric fields. This distortion inevitably affects the surrounding stress field. If these excited states propagate in one direction, the stress field developed around this movement will also propagate in the plane perpendicular to this direction. It is similar to the waves generated when a stone falls on a calm water surface. When we do stone skipping as shown in Figure 85, it is the electric current that these waves occur continuously, and the wave generated around the point of current is the magnetic field.

The BCS theory

In this regard, we first discuss the BCS theory, a quantum mechanical analysis of low temperature superconductivity. The BCS theory was proposed in 1957 by Bardin, Cooper, and Schrieffer of the United States, where BCS denotes the first alphabetic characters of these three names. The theory is known to explain the phenomena of low temperature superconductors (in short, LTSs, with T_c less than 30 K) composed mainly of metallic elements or alloys, but it is not reliable for HTSs composed mainly of oxides.

Figure 85. Waves formed by stone skipping (image from wikipedia.org). The movement of stone skipping (or the stone itself) is a current and the waves formed on the water surface are magnetic fields.

The BCS theory is a theory that explains superconductivity in terms of quantum mechanics.[179] As electrons move through a conductor, nearby cations in the lattice are deviated from their equilibrium positions under the influence of the electrons. This lattice distortion changes the charge distribution of cations, which in turn attracts electrons with the opposite sign of spin. This phenomenon is called the electron-phonon interaction, and electron pairs, called Cooper pairs, are formed via this interaction. This is not a phenomenon in which two electrons attract each other at the same time, but rather an interaction that requires time for the effect of one electron to be transferred to a cation and then to the other electron. The key feature of the theory is that these two electrons form a Cooper pair by the interaction with phonons, and the paired electrons conduct allegedly electricity without resistivity. Conventionally, the two electrons should have a repulsive force to repel each other by Coulomb's law. However, if temperature is sufficiently lowered, the two electrons are attracted to each other to form a Cooper pair, rather than repelling each other, and the paired electrons become more stable. The temperature at which this interaction occurs should be the critical temperature of superconductors, T_c. Cooper pairs in superconductors form one aggregated state. In order to decompose Cooper pairs below T_c, a large amount of energy is required because the entire aggregate are involved. At low temperatures below T_c, lattice vibrations are insufficient to break the electron pairs in the aggregate. The electrons do not move individually, but move as an aggregate, so that there is no resistance.

Superconductivity was observed even in thin films of atomic thickness.[180] if the Cooper pair were the cause of superconductivity, the BCS theory can not explain this kind of thin film superconductivity. Superconductivity in polymers[181] and buckyballs[182] has also been known for a long time, which can neither be explained by the theory.

This theory claims that a strong force that binds two electrons is required to explain high temperature superconductivity, but such a force has not yet found. The main reason why the BCS theory cannot be applied to HTSs is that it is based on the Landau quasiparticle concept, which cannot be clearly defined. In other words, the BCS theory starts from the mean field theory[*143)] of degree of freedom for the electrons. By combining electrons with phonons in the mean field, superconducting states can be achieved below T_c. In HTSs, this hierarchical separation of degree of freedom is not possible.[183] For this reason, the BCS theory is not enough to be fully relied upon. Substantially, a new paradigm for the electrical conductivity may be required. This book begins with the idea that the electrical conductivity in a conductor should be addressed in the same unified theory, regardless of whether the conductor is a low or high temperature superconductor or a normal conductor.

6.2. Superconductivity in the new paradigm

When describing superconductivity, there is no mention on the movement of protons or neutrons that make up atomic nuclei.[*144)] If one electron in an atom belongs to only one

---

[*143)] In physics, mean field theory deals with large and complex probability models using simpler models. To do this, a large number of interacting individual components are used. It replaces the influence of all other factors on one individual element with a single mean value, thus simplifying the problem of multiple systems into the problem of a single system.

[*144)] The BCS theory, of course, mentions the displacement of

proton, the change in the electron energy level must inevitably be accompanied by the proton movement (or the positional change). This cannot be inferred, if all the nucleons are concentrated at the atomic center. It was once controversial whether electrons existed in the nucleus due to the electrons generated during beta decay,*145) which raised doubts about the current atomic model. Of course, this problem was solved by the Swedish physicist Klein, who gave a quantum mechanical explanation denying the existence of electrons in the nucleus.*146) This explanation is simply a quantum mechanical calculation of whether or not electrons are contained in the atomic nucleus in the current atomic model, which is quite different from the new atomic model in which protons and electrons in atoms are paired in the form of hydrogen atom, called plectons, as introduced in Chapter 2.

In the regime of the new paradigm, plectons and neutrons are arranged to form a spherical atom as conceptually shown in Figure 20. Every atom looks like a grape cluster. The total energy of an atom varies depending on the locations of grape grains in the outmost shell, namely on the locations of valence plectons. One plecton accompanies almost always a neutron. We call this plecton-neutron pair PeN, as was previously introduced. When a valence PeN (a valence electron in current physics) moves to a new location on the atom surface relative to

---

cations.
*145) In negative beta decay, a neutron produces a proton and an electron and an electron antineutrino. When this reaction occurs in a nucleus, the element is converted to other element.
*146) See footnote 31.

other ones, the surrounding stress field developed in the solid vacuum changes, so the chemical bond energy changes. Therefore, as in the case of normal electrical conduction, it is also assumed that electricity flows via changing the locations of valence plectons and neutrons on the surface of the atom in a superconductor below T_c. We think that the current is not the movement of particles called electrons (or Cooper pairs in the BCS theory), but the movement of the potential energy of distorted ε_P bonds via the positional change of valence PeNs on the atom surface. Superconductivity is considered in terms of the movement of such a distorted state in this book.

Non-symmetric distortion around atoms

In the new paradigm, the (free) electrons in conventional physics is replaced by the distorted states of ε_P bonds. Energy is needed for valence PeNs to change their locations in response to external electric fields.*[147]) In an insulator, interatomic bonds between valence PeNs are rigid, the energy required for the deformation or movement of these chemical bonds is high, so the electrical conductivity is low. As temperature increases, the electrical conductivity of insulators increases because part of the activation energy to soften the rigid bonds can be supplied thermally. Since ε_P bonds (of a conductor) are flexible (or

---

*[147]) Atoms constituting a conductor thermally oscillate randomly, but when an external electric field is applied, this randomness disappears and the vibration becomes anisotropic. If the vibration of one atom is anisotropic, the vibration of adjacent atoms will also be affected and an anisotropy of vibrations will be formed in the corresponding directions.

because there are many such bonds), there are always randomly vibrating ε_P bonds. Electricity distinguishes negative and positive charges (it is anisotropic), but heat is essentially neutral (it is isotropic). In Planck's law, introduced in Chapter 5, heat generates neutral electromagnetic waves called radiation waves. Their wavelength and frequency depend on temperature and have the Planck distribution. Negative and positive charges combine to generate radiation, i.e, generate heat, which is equivalent to the combination of electrons and positrons to generate gamma rays. Resistant heat is thus generated via electrical shorts at the atomic level in the conductor.

But how do these micro electrical shorts occur? When the distorted states of ε_P bonds move in one direction, induced deformations of ε_P bonds called magnetic fields develop in the perpendicular plane. These induced deformations should again induce additional deformation of ε_P bonds with the orientation opposite to that of the primarily distorted ε_P bonds, and when these two antiparallelly aligned bonds are combined, they become neutral by releasing thermal vibration energy (ε_P bonds become isotropic and only thermally vibrating). This is the same situation as Faraday's law of Eq. (4.5), a magnetic field induced by the electric current in a conductor induces again an electromotive force in the direction to oppose the electric current, Since the induced electromotive force is generated in the interiors of a conductor, distorted ε_P bonds by the external electric field become neutral (electrically shorted) and thus heat is generated. This is the heat of resistance. In order to for a conductor to be a superconductor below T_c, the formation of induced secondary deformation of ε_P bonds (induced by the generated magnetic field) that eliminates primarily distorted ε_P bonds formed by the electric field must be suppressed. That is, the formation of magnetic fields within the conductor should be suppressed.

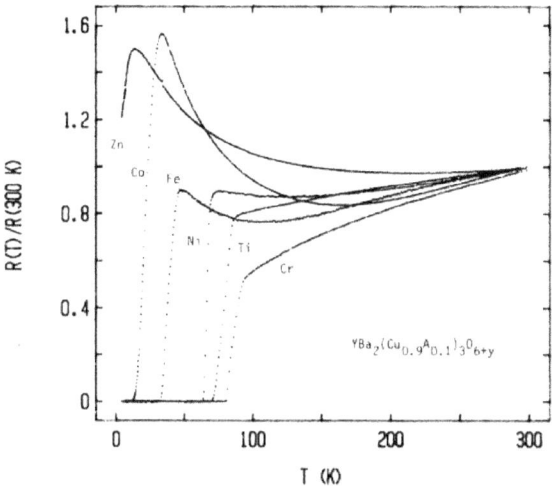

Figure 86. Some superconductors show rapid increases in the resistance upon approaching the critical temperature $YBa_2(Cu_{0.9}A_{0.1})_3O_{6+y}$, A=Ti, Cr, Fe, Ni, Zn (G. Xiao, et al.)

An electric current inevitably accompanies a magnetic field. As discussed in Chapter 5, a magnetic field in the new paradigm is an electromagnetic tensile or compressive stress field, appearing concentrically around the intersection of the plane perpendicular to the direction of a current, induced by the tension or contraction of ε_P bonds due to an external electrical stress field. The absence of magnetic field in a conductor means a state in which all the ε_P bonds are cross-linked to form a network structure so that the conductor is electromagnetically solidified. In this state, even if an external electric field is applied, the distortion of the network structure is not easy and the induced distortion called the magnetic field in the perpendicular plane to the electric field is not allowed. Accordingly the distortion of ε_P bonds is difficult or the stored energy in the distorted ε_P bonds cannot move, and thus the electrical resistance will become infinite. If no magnetic field is induced internally, as shown in Figure 83, there is a possibility that the resistivity increases rapidly as temperature decreases. Figure 86 illustrates this possibility. Some doped superconductors show rapidly increasing resistances as temperature is lowered to T_c.[184]

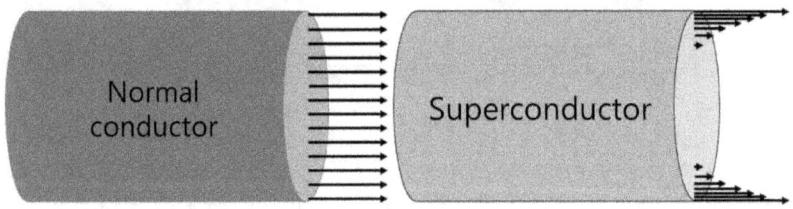

Figure 87. A normal conductor (left) carries electricity throughout the whole conductor, while a superconductor (right) carries electricity only on and near the surface. The surface thickness of superconductivity is called the London penetration depth.

However, because the solid vacuum near the free surface of a conductor can respond to the flow of electric field,*[148] electricity flows without resistance only through the surface

---

*148) Because magnetic fields can be induced in the solid vacuum by the distorted ε_P bonds in the surface layers of a superconductor.

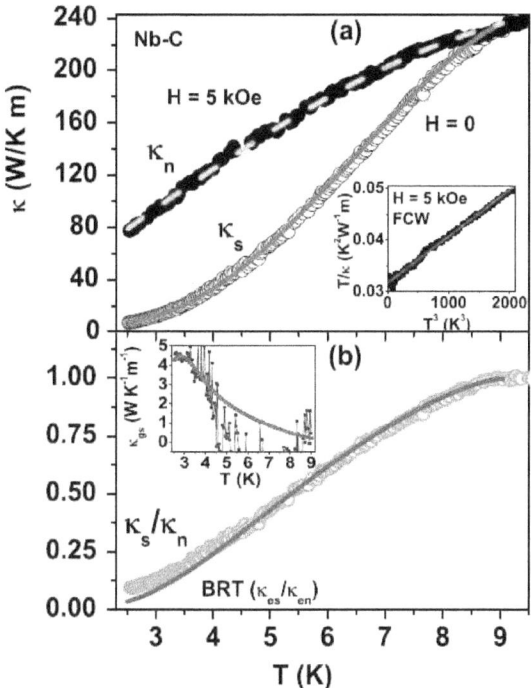

Figure 88. Difference between the thermal conductivity of Nb-C in the superconducting state and non-superconducting state in an external magnetic field: The thermal conductivity is significantly reduced in the superconducting state. (a) temperature dependence of the thermal conductivity, (b) its ratio of the two states (L.S. Chandra, et al.)

layers, as shown in Figure 87. A superconductor is paradoxically a perfect insulator below T_c, but it becomes a superconductor via interaction with the solid vacuum. This also means that below T_c, thermal activation in superconductors is difficult. Since radiant heat is a kind of electromagnetic waves, it is thus expected that the thermal conductivity will also be low if the formation of magnetic

Table 6. Periodic Table of Superconductivity (wikipedia.org)

H		Atmospheric pressure superconductor		Only high pressure superconductor							B	C	N	O	F	He
Li 0.0004 14 30	Be 0.026 3.7 30		T_c(K) T_c^{max}(K) P(GPa)		T_c^{max}(K) P(GPa)						B 11 250	C	N	O 0.6 100	F	Ne
Na	Mg										Al 1.14	Si 8.2	P 13 30	S 17.3 190	Cl	Ar
K	Ca	Sc	Ti 0.39 29 217	V 5.38 19.6 106	Cr 3.35 56.0	Mn 16.5 120	Fe 2.1 21	Co	Ni	Cu	Zn 0.875 7 1.4	Ga 1.091 5.35 11.5	Ge 2.4 32	As 8 150	Se 1.4 100	Br	Kr
Rb	Sr	Y	Zr 0.546 7 50	Nb 9.50 19.5 115	Mo 0.92 11 30	Tc 7.77 9.9 10	Ru 0.51	Rh .00033	Pd	Ag	Cd 0.56	In 3.404 5.3 11.3	Sn 3.722 3.9 25	Sb 7.5 35	Te 1.2 25	I	Xe
Cs 1.3 12	Ba 5 18	La- Lu	Hf 0.12 8.6 62	Ta 4.483 4.5 43	W 0.012	Re 1.4	Os 0.065	Ir 0.14	Pt	Au	Hgα 4.153	Tl 2.39	Pb 7.193 8.5 9.1	Bi	Po	At	Rn
Fr	Ra	Ac- Lr	Rf	Ha													

La 6 13 15	Ce 1.7 5	Pr	Nd	Pm	Sm	Eu 2.75 142	Gd	Tb	Dy	Ho	Er	Tm	Yb 12.4 174	Lu
Ac	Th 1.368	Pa 1.4	U 0.8(β) 2.4(α) 1.2	Np	Pu	Am 0.79 2.2 6	Cm	Bk	Cf	Es	Fm	Md	No	Lr

field is suppressed in superconductors. An example is shown in Figure 88.[185] The thermal conductivity decreases as temperature decreases, but the thermal conductivity of the superconductor decreases more rapidly. In general, the higher the electrical conductivity of a conductor, the higher the thermal conductivity. In the case of superconductors, this trend is reversed.

Consequently, superconductivity is due to the interaction

of the internal structure of superconductors near the surface with the solid vacuum under the influence of external electric fields. So why do the individual metals in Table 6[186] differ in T_c? Consider some examples of metal elements. The key to this consideration is what crystal structures a conductor can easily form ε_P bonds and how high the bond strength will be.

Superconductivity of metallic elements

Metals with one valence PeN, such as lithium (Li), cannot form enough ε_P bonds because the PeN is fully used for primary chemical bonds (or there may be no distinction between the primary and ε_P bonds). Therefore, most Group 1 metallic elements are unlikely to exhibit superconductivity. However, the fact that superconductivity occurs at very low temperatures may be considered regarding to the subatomic helium structure [He] which is not completely spherical. Li exhibits superconductivity at very low temperatures (below 0.0004 K) and has a relatively high T_c of 14 K at 30 GPa, as shown in Table 6. The superconductivity of Li may be related to the behavior and morphology of He, which liquefies at 4 K.

The subatomic structure of sodium (Na), [neon (Ne)], is more spherical than [He]. Superconductivity does not occur even at high pressures. Beryllium (Be) has a higher T_c (0.026 K) because $2s^2$ PeN can be involved in primary and ε_P bonds and therefore [He] can be more closely accessed. On the other hand, [He] is more difficult to be accessed at high pressures due to the two primary bonds, and thus T_c is lower (3.7 K) than that of Li at high pressures.

Table 7. Pressure-dependent crystal structure of Si (M. Debessai, et al.)

	Crystal structure	Pressure Range (GPa)
I	cubic (diamond structure)	0 → ~11
II	body-centered tetragonal (β-Sn)	~11 → 15
III	body-centered cubic	~10 → 0
V	primitive hexagonal	~14 → 40
VII	hexagonal close-packed	~40

Magnesium (Mg) has the same hcp structure as Be, with two $3s^2$ PeNs participating in the 12 bonds. The number of ε_P bonds is not enough for superconductivity to appear.

T_c of Aluminum (Al) is 1.14 K at atmospheric pressure, much higher than that of Li. Three PeNs ($3s^2\ 3p^1$) may be involved in chemical bonding. Al has a fcc structure with the coordination number (the number of nearest atoms) of 12, like Mg with a hcp crystal structure. If two PeNs are responsible for the 12 bonds, $3p^1$ PeN can form ε_P bonds, resulting in superconductivity at 1.14 K.

In silicon (Si), $3s^2\ 3p^2$ PeNs may be involved in the bond, and in fact, all four PeNs participate in primary bonds to form a diamond structure. This results in a lower density than Al, but a much higher melting point. The absence of ε_P bonds does not allow superconductivity at atmospheric pressure. However, the crystallographic structure is distorted (a phase transition) at high pressures,[187] leading to the occurrence of superconductivity.[188] Comparing Table

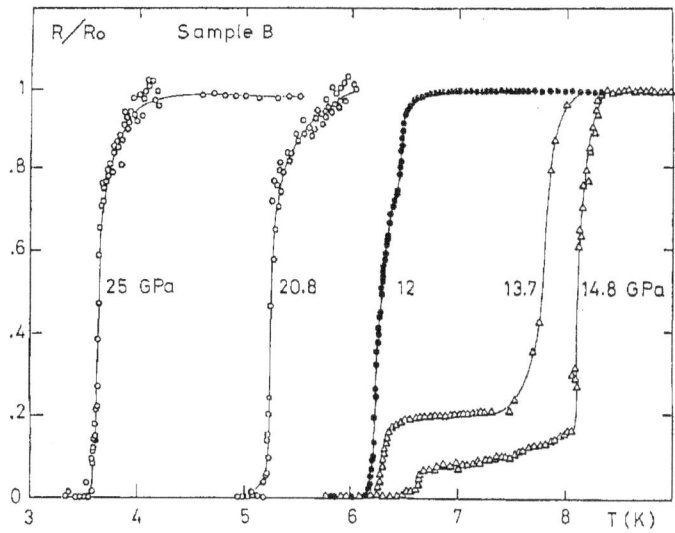

Figure 89. Superconductivity of Si depending on the pressure (G Martinez, et al.)

7 and Figure 89, it is noticed that T_c of Si does not increase consistently with pressure, but it is the highest at the pressure range at which a specific crystal structure (body-centered tetragonal) is formed. Lead (Pb), which has the same valence PeN number as Si, has a fcc crystal structure and T_c of 7.193 K, which is quite high. The fact suggests that ε_P bonds by $6p^2$ PeN are clearly present and the bond strength would be high.

Copper (Cu) does not exhibit superconductivity. There are three Cu oxides: CuO, CuO_2, and Cu_2O. In other words, Cu may have 1, 2, or 4 PeNs participating in primary chemical bonds. Since the crystallographic structure of Cu is fcc, if two valence PeNs participate in primary bonds, the other two may participate in ε_P bonds. The PeN configuration of

Cu is [Ar] $3d^{10}\ 4s^1$. In order for superconductivity to occur, some $3d^{10}$ PeNs must be involved in ε_P bonds. As Cu does not allow superconductivity, it is thus assumed that none of $3d^{10}$ PeNs are involved to the bonding or its binding energy is very low.

Mechanism of superconductivity

Below the critical temperature of a superconductor, ε_P bonds are cross-linked and an electromagnetically solid structure is formed (an electromagnetic phase transformation from liquid to solid) and only ε_P bonds near to the surface can be individually distorted in response to external electric fields.*[149]) An internal electric field is created by these bonds only at the surface and a magnetic field is induced in the nearby solid vacuum. Thermal energy is not enough to break the whole network of ε_P bonds in the interiors of the conductor below T_c, and thus no internal magnetic fields are allowed. Electricity flows only through the surface layers. The magnetic field formed in the solid vacuum causes zero resistance because it does not induce the opposite electromotive force (electromagnetically, the solid vacuum is a perfect elastomer).

The BCS theory states that aggregated Cooper pairs in superconductors flow undisturbed and without resistance below T_c. These electron pairs, in the regime of the new paradigm, are analogous to the pairs of valence PeNs participating in ε_P bonding. This is just one of the

---

*[149]) The surface layer has a more flexible bond structure than the interiors of the conductor.

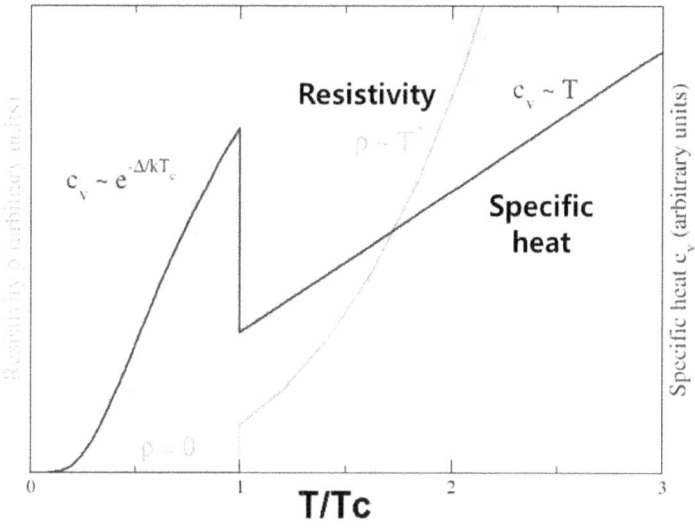

Figure 90. Variation of the heat capacity and resistivity near T_c

conditions for the flow of electricity in the new paradigm. A valence PeN of an atom in a ε_P bond is excited by an external electric field and reacts with another valence PeN of the adjacent atom so that the PeN configuration of the atoms is modified. Electricity flows, as these modified bonding structure moves in the direction of the external electric field.*150) The electrical energy can transferred because there are always thermal vibrations moving randomly in a wave form. When an external electric field is applied, such random vibrations become anisotropic which propagate as a wave. The supercurrent by Cooper pairs of the BCS theory is nothing but the flow of the deformed states of ε_P bonds.

---

*150) Actually, only the relative displacement changes periodically in the form of wave.

Figure 91. Unit cell of the A15 phase of Nb_3Sn

On the other hand, below T_c, ε_P bonds cannot be thermally excited in superconductors. This is because the thermal excitation of ε_P bonds is also an electromagnetic phenomenon. Since magnetic fields can be only formed in the solid vacuum near the conductor surface, the nearby ε_P bonds can respond to the external electric field. Note the variation of heat capacity and resistivity of superconductors near T_c in Figure 90. The heat capacity varies discontinuously at T_c. It is a phase transformation in which ε_P bonds are involved. Below T_c, the whole network of ε_P bonds thermally vibrating in the electromagnetically solidified state. Therefore, the heat capacity below T_c is completely different from that above T_c.

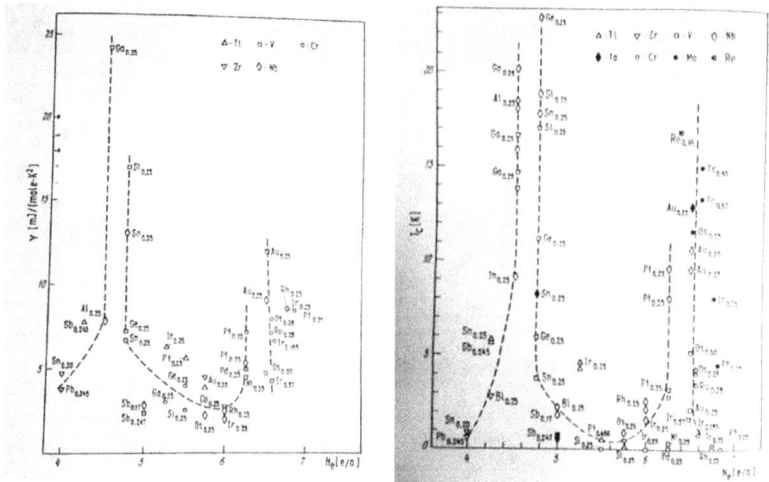

Figure 92. Electron specific heat γ and superconductivity critical temperature T_c according to the number of valence PeN of A_3B alloys (Vonsovsky, Izyumov, Kurmaev).

As shown in Figure 92, the variation of the electron specific heat and T_c as a function of the number of valence PeN of A_3B alloy with the A15 phase structure[*151] are almost similar.[189] This fact suggests that the more valence PeNs that do not participate in primary bonds, the higher the critical temperature.

What are the factors that determine T_c for low temperature superconductors (LTSs)? If the energy necessary for destroying the ε_P bond network formed below T_c is high, T_c will be high (the bond strength must be

---

[*151] A15 (β-W or Cr_3Si structure) is an intermetallic compound with the chemical formula A_3B (A is a transition metal and B can be any element) and has a crystal structure as shown in Figure 91. It has a relatively high T_c of ~20 K and maintains superconductivity even at tens of tesla. This is called type II superconductivity.

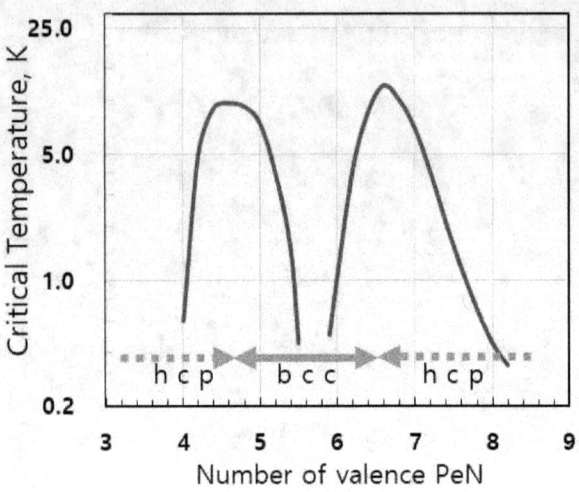

Figure 93. The critical temperatures of the transition metal alloys are high between the two crystal structures (Vonsovsky, Izyumov, Kurmaev).

high). The binding energy will be higher if there are two or more valence PeNs available for an atom. Also, if PeNs for ε_P bonding are provided from the PeNs in the primary bonds via crystallographic structural transformation, T_c will increase. High pressures may provide such a condition, so LTSs have usually the lower critictal temperatures at higher pressures. Referring to Figure 93,[190] transition metals whose crystal structure is somewhere between bcc-fcc and bcc-hcp have high critical temperatures. This means that the crystal structure based on primary bonds is unstable and a large number of PeNs can participate in ε_P bonds. Another consideration is the range of interatomic distance for ε_P bonds. The shorter this distance, the higher the likelihood of ε_P bonding, and the lower the critical temperature for many metals. However, as in the case of Si in Table 7 and Figure 89, the pressure-dependency of T_c in metals or alloys does not indicate that T_c varies consistently with pressure, namely the interatomic distance.

In summary, T_c is high when the number of available PeNs participating in ε_P bonds is higher and thus the critical bond density for the electromagnetic solidification is reached at higher temperatures. If the crystal structure formed based on primary bonds is unstable, the number of available PeNs for ε_P bonds increases. Also, if the strength of ε_P bond is high, high T_c will be obtained because the energy to break this bond is high. When pressurized, the average ε_P bond length is reduced so that the ε_P bond strength could becomes higher, leading to a higher T_c. The distortion of the crystal lattice may lead to increase or decrease the number of available PeNs for ε_P bonds, so the effect of pressure on T_c is not consistent.

6.3. High temperature superconductors

High temperature superconductors (HTSs) have low electrical conductivity at room temperature. The minimum resistivity near the critical temperature T_c of a HTS in Figure 94 is 100 Ωcm, far higher than those of typical conductors in Table 4. Unlike metallic Type I superconductors (low temperature superconductors, LTSs), HTSs exhibit superconductivity at high temperatures (above 77 K, the boiling point of nitrogen), and their electrical conductivity at room temperature is closer to those of

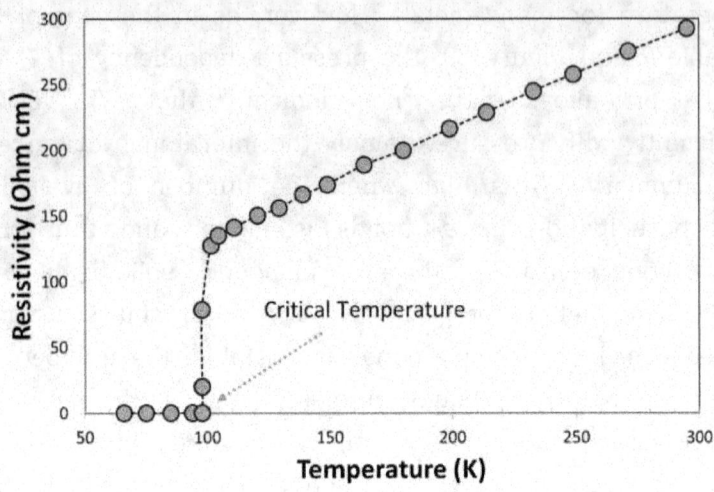

Figure 94. Resistivity of a HTS as a function of temperature.

insulators than to metallic conductors. In addition, superconductivity does not discontinuously disappear at a certain magnetic field, and magnetic fields penetrate into the superconductors within a magnetic field strength range. This kind of superconductors is referred to as Type II.[191]

II-type superconductors

Type II superconductors form magnetic vortices above a certain magnetic field strength, $H = H_{c1}$, the primary critical magnetic field strength, as shown in Figure 95.[192] The vortex density increases as H increases. If H is higher than H_{c2}, the secondary critical magnetic field strength, superconductivity disappears as shown in Figure 96. Thus, magnetic fields are not completely ousted from Type II superconductors, and the Meissner effect is limited.[193]

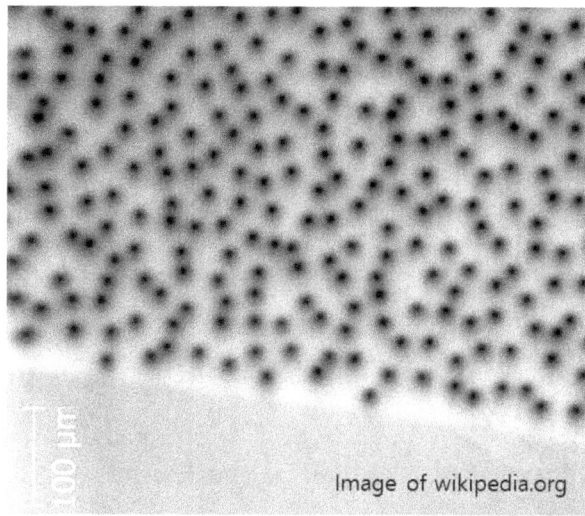

Figure 95. Magnetic vortices (spots) observed on the thin films of a YBCO HTS.

Magnetic vortices, also known as Abrikosov*152) vortices, are not in the superconducting zone, but is surrounded by rotating supercurrents in the superconducting zone. Abrikosov vortices form a lattice with some regularity, as shown in Figure 95.[194] With magnetic vortices, flux pinning is possible in which the superconductor is held in a space above a permanent magnet. Flux pinning by superconductivity can be used for elevators, frictionless bonding, magnetic levitation trains, and the like. The thinner the superconductor, the stronger the flux pinning.

---

*152) Alexei A. Abrikosov (6.1928 - 3.2017) was a Russian and American theoretical physicist. He has made many contributions to solid state physics. In 2003 together with V.L Ginzburg and A.J. Leggett, he won the Nobel Prize in Physics for the theory of material behaviors at cryogenic temperatures.

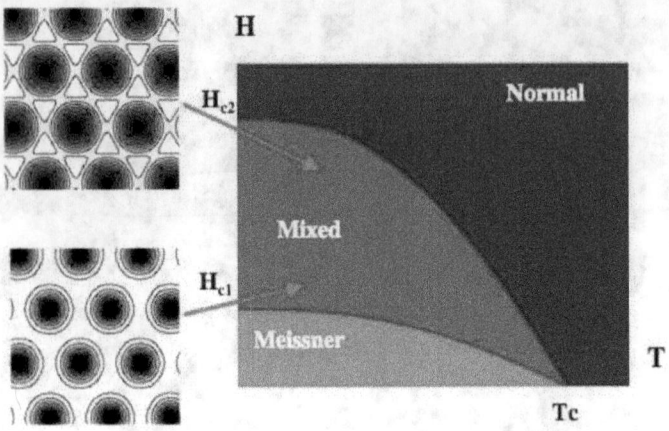

Figure 96. Behaviors of type-II superconductors (by X. Yu, "Vortices in Type-II Superconductors" @guava.physics.uiuc.edu)

Various kinds of HTS

The first HTS was discovered by Bednorz and Müller*[153]) in 1986.[195] Superconductivity occurs when cooled with liquid nitrogen because T_c for some recently found HTSs is higher than the boiling point of nitrogen, 77.2 K (-195.8°C). Until 2008, only copper oxide ("cuprate") based HTSs were known and HTSs meant copper oxide HTSs. Since then, high temperature superconductivity was also found in iron compounds.[196] In 2015, hydrogen sulfide (H_2S) under ultrahigh pressures (~ 150 GPa) showed superconductivity at about 203 K (-70°C), in which H_2S transforms to H_3S. This superconductor has the highest T_c known to date.[197]

---

*[153]) Johannes G. Bednorz (5.1950 -) was a German physicist. He was awarded the Nobel Prize in Physics in 1987 with K. Alex Müller (4.1927-, Swiss physicist) for the discovery of HTSs.

The copper oxide superconductor has a multi-layered perovskite structure in which the crystal structure is distorted and lacks oxygen. Superconductivity occurs between the multiple copper dioxide (CuO_2) layers. Owing to this structure, electrical conductivity is highly anisotropic, and the more CuO_2 layers, the higher the critical temperatures. Since electricity flows by electron holes[*154] of the oxygen atoms of CuO_2, the electrical conductivity in the plane parallel to the CuO_2 layer is much higher than in the perpendicular directions.

The first superconductor with T_c above 77.2 K, the liquid nitrogen boiling point, is $YBa_2Cu_3O_{7-x}$ and is also referred to as YBCO123 or simply YBCO. In the unit cell of $YBa_2Cu_3O_7$ three perovskite unit cells overlap. Referring to Figure 97, each unit cell has a Y or Ba atom in its center. Ba is in the bottom lattice, Y is in the middle lattice, and Ba is again in the top lattice. Therefore, Y and Ba are in the [Ba-Y-Ba] array along the c-axis (vertical axis). The unit cell edge is occupied by two types of Cu ions, Cu^{+1} and Cu^{+2}, depending on the degree of oxidation. Oxygen atoms occupy the four locations O(1), O(2), O(3) and O(4).[198] The three perovskite unit cells can have nine oxygen atoms, but in reality $YBa_2Cu_3O_7$ has seven oxygen atoms, which make an oxygen-deficient perovskite structure. The stacking order is as follows.

$$(CuO)(BaO)(CuO_2)(Y)(CuO_2)(BaO)(CuO)$$

[*154] An electron hole is a PeN-deficient state, hence a bond is lacking. The lacking bond between Cu and O appears to respond to external electric fields. This means that there are many available valence PeNs of Cu for ε_P bonds.

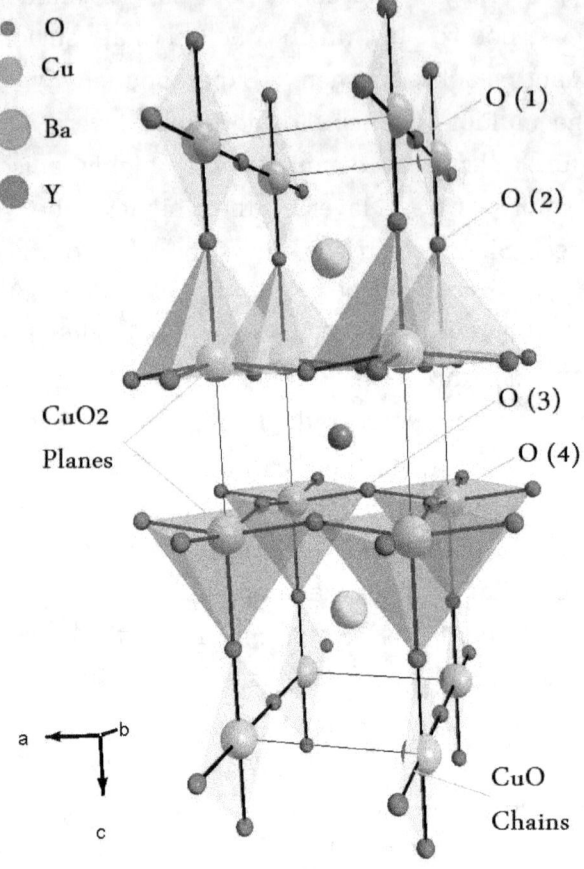

Figure 97. YBCO unit cell (image from wikipedia.org)

In this superconductor, the Y atom separates the two CuO_2 layers. T_c is the highest as 95 K at x ~ 0.07, and its crystal structure is orthorhombic.*155)199 At x ~ 0.6, the structure converts to tetragonal and loses superconductivity.200

---

*155) It is a crystal system in the shape of a square pillar with a rectangle as the basal plane, and the length of both sides of the basal plane is different from each other.

In copper oxide superconductors, T_c varies depending on the doping as shown in Figure 98. Without doping, the phase is antiferromagnetic (AF) with the highest Curie temperature.[201] YBCO can be generally regarded as a two-dimensional material with regard to the electrical conductivity, and the superconductivity is determined by the behavior of electrons in the weakly bound CuO_2 plural layers. The copper oxide layer is shaped like a chessboard with O_2^- ions at the corners of the square and Cu_2^+ ions at the center. The unit cell forms an angle of 45° with this square. Adjacent layers with ions such as lanthanum (La), barium (Ba) and strontium (Sr) stabilize the structure and provide electrons and holes in the copper oxide layer.

Bi-, Tl- and Hg-based HTSs have similar crystal structures.[202] Like YBCO, they contain perovskite unit cells and CuO_2 layers, but, unlike YBCO, no Cu-O chain. While YBCO superconductors are orthorhombic, they are tetragonal.

Bi-based superconductors have chemical formulae of $Bi_2Sr_2Ca_{n-1}Cu_nO_{4+2n+x}$ (BSCCO), n = 1, 2, and 3. They are classified into three types, Bi-2201, Bi-2212 and Bi-2223, with T_c of 20, 85 and 110 K, respectively.[203] Two of them are tetragonal and consist of two oblique unit cells with a Bi-O bilayer, where the Bi atom in one layer is located below the O atoms in the adjacent layer. The Ca atoms form one side in the middle of the CuO_2 layer in Bi-2212 and Bi-2223. Bi-2201 has no Ca. Bi-2201, Bi-2212 and Bi-2223 have one, two and three CuO_2 layers, respectively. When there are many CuO_2 layers, the c-axis lattice constant becomes large. The coordination number*[156] of Cu is also different.

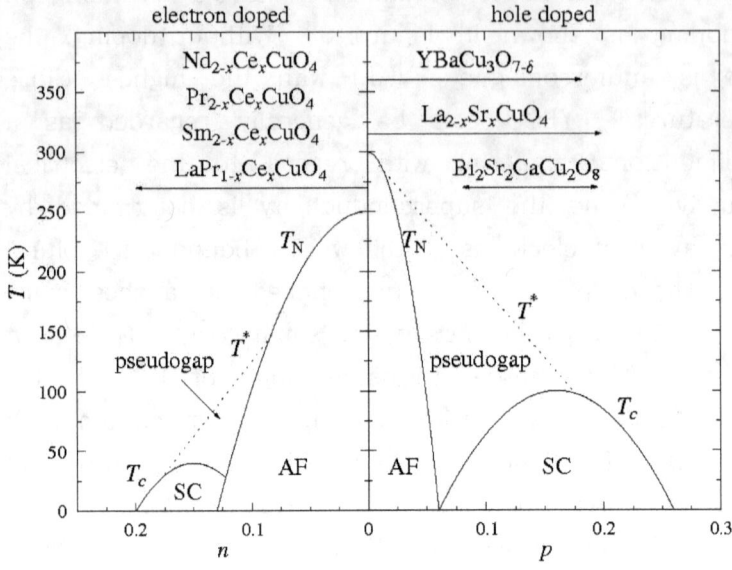

Figure 98. Effect of electron (n) and hole (p) doping on T_c in copper oxide superconductors (C. Hartinger).

In Bi-2201, Cu has an octahedral coordination number with respect to O. Bi-2212 has a pyramidal structure, in which the Cu atom is surrounded by five O atoms. In Bi-2223, the Cu atom with coordination number 2 combines with four O atoms to form a square plane, and the other atom has a pyramid structure with five O atoms.[204]

The first Tl-based oxide superconductor (TBCCO) has one Tl-O layer for $TlBa_2Ca_{n-1}Cu_nO_{2n+3}$, and two layers for $Tl_2Ba_2Ca_{n-1}Cu_nO_{2n+4}$ with n = 1, 2 and 3.[205] $Tl_2Ba_2CuO_6$ is called Tl-2201 and has one CuO_2 layer, and has a laminated structure as follows.

---

*156) The number of closest atoms surrounding (or coordinating) an atom.

(Tl-O)(Tl-O)(Ba-O)(Cu-O)(Ba-O)(Tl-O)(Tl-O)

$Tl_2Ba_2CaCu_2O_8$, i.e. Tl-2212, has two CuO_2 layers with a Ca layer in between. Similar to Tl-2201, the Tl-O layer is outside the Ba-O layer. Tl-2223 ($Tl_2Ba_2Ca_2Cu_3O_{10}$) has three CuO_2 layers sandwiching two Ca layers. In Tl-based superconductors, T_c is high when the number of CuO_2 layers is large. However, in $Tl_2Ba_2Ca_{n-1}Cu_nO_{2n+3}$, T_c is lowered for n > 3, and in $Tl_2Ba_2Ca_{n-1}Cu_nO_{2n+4}$, for n > 2.[206]

Hg-based superconductors include Hg-1201 ($HgBa_2CuO_4$), Hg-1212 ($HgBa_2CaCu_2O_6$) and Hg-1223 ($HgBa_2Ca_2Cu_3O_8$) and their crystal structures are similar to those of Tl-1201, Tl-1212 and Tl-1223. Only Hg is substituted for Tl. T_c of Hg-1201 with one CuO_2 layer is much higher than Tl-1201. In Hg-based superconductors, T_c increases with the number of CuO_2 layer. The critical temperatures of Hg-1201, Hg-1212 and Hg-1223 are respectively 94, 128 and 134 K.[207] In Hg-1223, T_c goes up to 153 K at high pressures and is sensitive to the crystal structure of the compound.[208]

Fe-based superconductors contain Fe and Group 5A (As or P) elements or Group 6A elements in the periodic table of elements. T_c is the highest after the copper oxide system. The highest critical temperature reported in 2014 was above 100 K for FeSe films.[209] LnFeAs (O, F) or $LnFeAsO_{1-x}$ (Ln = a lanthanide element) have T_c of ~56 K and are called 1111 compounds.[210] $(Ba,K)Fe_2As_2$ superconductors have Fe-As layers in pairs and are called 122 compounds with T_c of 38 K.[211] 111 compounds LiFeAs and NaFeAs have T_c of up to 20 K.[212]

Most of undoped Fe-based superconductors are magnetically aligned at low temperatures after the tetragonal-orthorhombic phase transformation, similarly to copper oxide superconductors.[213] However, they are more like metals than Mott*[157] insulators*[158] and have five energy bands on the Fermi surface.*[159)214] As shown in Figure 99, the phase diagrams of doping in Fe-based superconductors containing As are very similar regardless of the constituent elements (La[215], Sm[216], Ce.[217], Ba-122[218]). The superconducting phase adjoins or overlaps with the region of the magnetic phase. T_c shall vary with the As-Fe-As bond angle with the highest one in the untwisted FeAs$_4$ tetrahedron.[219]

---

*157) Nevill Francis Mott (9.1905 - 8.1996) was a British physicist. He was awarded in 1977 the Nobel Prize in Physics with P.W. Anderson and J.H. Van Vleck for the study of the electronic structure of magnetic materials and amorphous semiconductors. Mott and Anderson have found why magnetic and amorphous materials are sometimes conductors and sometimes insulators.

*158) According to band theory, a Mott insulator should be a conductor, but it is actually an insulator. This is due to the electron-electron interaction that was not covered in previous theories. Band gaps in Mott insulators exist between the bands of similar properties, such as the $3d$ properties. This is different from the insulator band gap in the anion and cation states, such as between the O $2p$ and Ni $3d$ bands in NiO.

*159) The Fermi surface is the surface in the reciprocal lattice space and distinguishes between the filled and unfilled electron energy levels at 0 K. The shape is determined by the symmetry of unit cell of the crystal and the filled electron energy band. The Fermi surface is resulted directly from Pauli's exclusion principle, which allows only up to one electron in a quantum state.

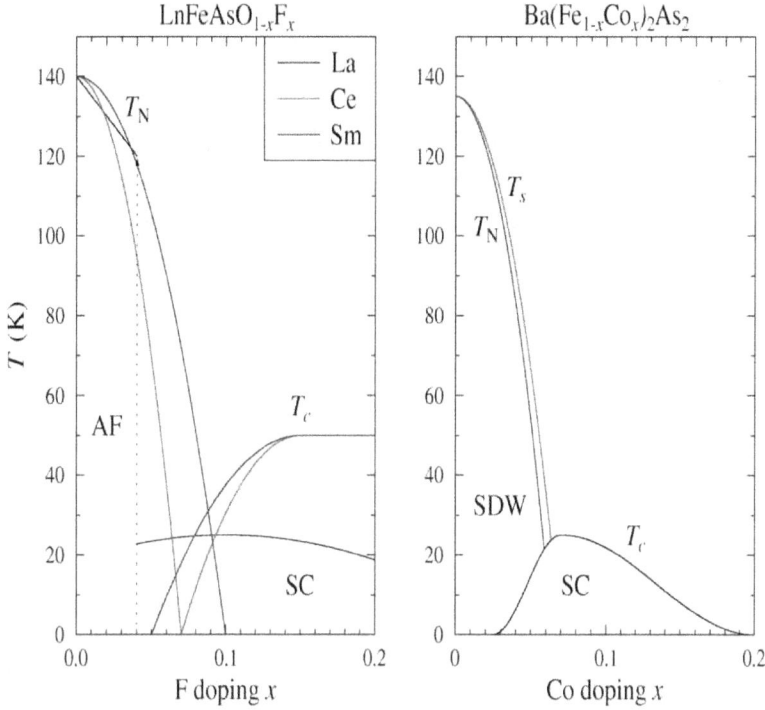

Figure 99. Phase diagram as a function of doping in Fe-based superconductors Ln-1111 and Ba-122 (image from wikipedia.org).

At pressures above 90 GPa, hydrogen sulfide becomes a metallic conductor, and as pressure increases, T_c increases to 23 K at 100 GPa and 150 K at 200 GPa, respectively.[220] When pressurized at high temperatures and cooled, it rises to 203 K (-70° C). the highest as of 2015. It is also predicted that if sulfur is replaced with phosphorus and further pressurized, superconductivity could occur even at room temperature above 0°C.[221]

Mechanism of high temperature superconductivity

Although the mechanism of high temperature superconductivity is unclear, there are common features.222 Low temperature antiferromagnetism in undoped materials and the occurrence of superconductivity by doping, the orbital states of Cu^{2+} ions are mainly d_{x2-y2} (see Figure 100), and the electron-electron interaction is considered to be more important than that of electron-phonon. These are interpreted as different features from low temperature superconductivity. Studies on the Fermi surface show that the overlap occurs at four points in the diamagnetic Brillouin[160] zone[161] where spin waves are present and the superconducting energy gap is higher at these points. The isotope effect is weak in most copper oxide superconductors, which cannot be explained by the BCS theory.

In undoped HTSs, T_c is inversely proportional to the square of the London penetration depth λ_L,[162] and the

---

[160] Léon Nicolas Brillouin (1889.8 - 1969.10) was a French physicist. He contributed to the development of solid-state physics such as quantum mechanics and radio wave propagation in air.

[161] The Brillouin zone is a unit cell defined only in the reciprocal lattice space. The outer surface of this unit cell consists of the planes associated with points on the reciprocal lattice. The Brillouin zone is important for describing the behavior of waves in crystals with repeating structures. The first Brillouin zone is the location of the points closer to the origin of the reciprocal lattice. Although the wider second and third zones are defined, the first zone is mainly used and the Brillouin zone also means the first zone.

[162] λ_L is the depth of magnetic field penetrating the superconductor. It is the depth at which the magnetic field

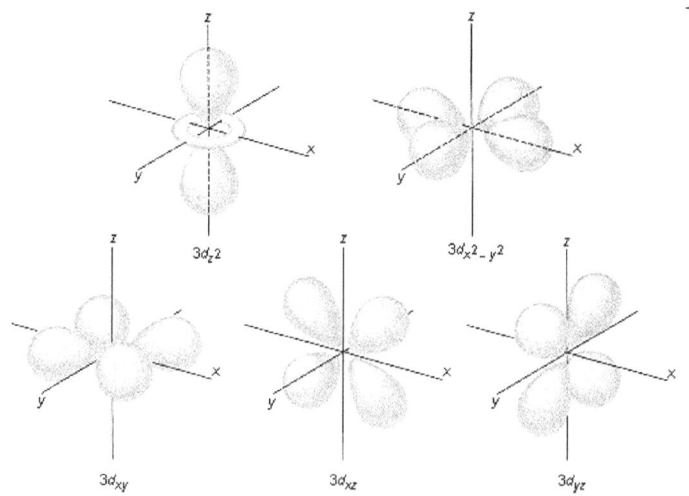

Figure 100. The five $3d$ orbitals : d_{xy}, d_{yz}, d_{xz}, $d_{x^2-y^2}$, d_{z^2} (wikipedia.org).

proportionality constants differ for hole and electron doping. This means that superconductivity in these materials is two-dimensional. The Nernst effect*163) is

---

strength is e^{-1} times of that at the superconductor surface. It is usually about 50 to 500 nm. The magnetic field strength at the surface is reduced by the ratio exp $(-x/\lambda_L)$ at the depth x from the surface. λ_L is determined from the supercurrent density which is closely related to T_c in HTSs. λ_L can be measured with a muon spin spectrometer, where the superconductor must not have its own magnetic structure. It can be obtained from the speed at which the polarization by muon spin disappears.

*163) The Nernst effect is a type of thermoelectric phenomenon in which an electric field is formed in the direction perpendicular to the direction of the magnetic field and the temperature gradient when a conductor is in a magnetic field and a temperature gradient is formed in the direction perpendicular to the magnetic field. Electrons in the conduction band in a semiconductor are statistically affected by temperature due to

evident in the phase of superconductivity or of pseudogap. A pseudogap refers to an energy gap formed only partially on the Fermi surface. It is the band structure at a specific point where there is a gap on the Fermi surface.[223] This gap is formed by the interaction of electrons with the lattice. The pseudogap is similar to the actual energy gap and there is no allowed energy level in between. Gaps refer to insulating states, which exist when electrons move parallel to the Cu-O bonds.[224] At 45° angle, the electrons move freely. The Fermi surface thus consists of the Fermi arcs that form pockets at the corners of the Brillouin zone. In the pseudogap, these arcs gradually disappear as temperature decreases, with the exception of only four points on the diagonal of the Brillouin zone. This means a completely new electronic phase where there is no possible energy state and electrons are superconducting in pairs. In addition, the presence of pseudogap may mean from the similarity between this partial gap and the superconducting gap that electrons formed Cooper pairs.

The electron configuration (i.e. that of PeN) of copper oxide superconductors is highly anisotropic. The Fermi surface is thus very close to the doped CuO_2 plane and is in the two-dimensional reciprocal space (or momentum = frequency space) of the CuO_2 lattice. The typical Fermi surface in the primary CuO_2 Brillouin zone is shown in Figure 101 (left). When doped with holes (p), as shown in Figure 98, superconductivity is usually observed, with the

---

an increase in the kinetic energy with temperature. If a magnetic field exists perpendicular to the temperature gradient, the charged particles including electrons feel a force in the direction of their movement and in the direction perpendicular to the magnetic field. This produce an electric field.

Fermi surface being the same as that of holes (opened as in Figure 101), resulting in a two-dimensional anisotropy, a unique electronic characteristic of HTSs.

According to a weak-coupling theory,*[164] superconductivity comes from antiferromagnetic spin fluctuations in the doped state.[225] If the wave functions in the copper oxide superconductor are paired, they should have a $d_{x^2-y^2}$ symmetry. Therefore, it is very important first to verify that the paired wave functions have a d-wave symmetry in order to confirm the spin fluctuation mechanism. Second, according to the interlayer coupling model, superconductivity can be enhanced in HTSs through BCS-type (s-wave symmetry) layers.[226] Adding the tunneling effect between the layers, the model can account for high temperature superconductivity as well as the anisotropic symmetry of the order parameters.*[165] Some d symmetry was confirmed by measurements of photoemission spectroscopy (PES),*[166] NMR, and specific heat, but a feature for the s symmetry was also confirmed and lacks consistency.

The copper dioxide layer is an insulator. It becomes a conductor by doping, and by adjusting the doping concentration T_c can be maximized. La_2CuO_4 is an insulator in which CuO_2 and LaO layers are alternately repeated. If

---

*[164] A weak-coupling theory is similar to the concept of ε_P bonds in the new paradigm of electrical conduction.

*[165] A property difference between two phases in the transition from one state to another. The order parameter is the difference in the density when a liquid changes into a gas.

*[166] A PES device analyzes the energy of electrons emitted from solids, liquids and gases using the photoelectric effect, to determine the electron binding energy at the substance.

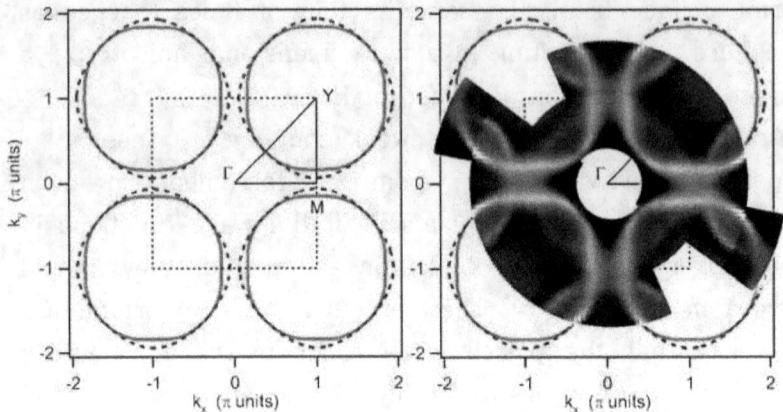

Figure 101. The Fermi surface of bi-layer BSCCO, calculated (left) and measured by ARPES (right). The dashed rectangle represents the first Brillouin zone (image from wikipedia.org).

8% of La is replaced with Sr by doping, Sr mediates interlayer bonding, and the CuO_2 layer becomes a conductor. For this reason, the fundamental interaction of superconductivity is sometimes argued to be the interaction of electron-phonon as with Cooper pairs. The undoped material is antiferromagnetic, whereas a few percent doping yields a small energy pseudo gap in the CuO_2 layer, but this is argued also due to phonon. As the number of charge carriers increases, this gap decreases to approach the superconducting gap. High T_c is therefore due to the percolating behavior of charge carriers. They move zigzag mainly in the conducting zone of the CuO_2 layer and, if doped, move to the adjacent conducting zone. The flexible bonding of the main lattice leads to a strong electron-phonon interaction in the interlayer dopants, which leads to high critical temperatures.[227]

The symmetry of the order parameter is best checked when Cooper pairs cross Josephson junctions[*167)] or have weak links.[228] Although spontaneous magnetization was expected to occur only at the contacts of the d symmetry superconductors, the experimental results were uncertain. This is believed to be due to impurities in the superconductor, and experiments have been devised to estimate such effects.[229] Although spontaneous magnetization in YBCO was clearly observed in these experiments, they showed the d symmetry of the order parameter. However, YBCO is tetragonal, so some s symmetry may be included. Subsequent experiments confirmed that the s symmetry was 3%.[230] In addition, pure d_{x2-y2} order parameters were found in orthorhombic $Tl_2Ba_2CuO_6$.[231]

The mechanism of high temperature superconductivity is still ambiguous. Most theoretical calculations, however, point out that the interaction between electrons is due to

---

[*167)] Brian David Josephson (1.1940-) was a Wales physicist and professor of physics at the University of Cambridge. In 1973, he received the Nobel Prize in Physics for the theoretical proof of the Josephson effect. This theory was established when he was a 22-year-old doctoral student. Josephson junction refers to a device in which the Josephson effect appears. The device consists of several superconductors with an insulator film between two superconducting layers. The absence of an external voltage also produces a supercurrent cross the device, the Josephson effect. This effect is a phenomenon of macroscopic quantum mechanics predicted by Josephson in 1962. This effect was observed before 1962, but it was thought that a "super short" caused the insulator film to be broken and electrons migrate. Previously, electrons in the conductor could pass through the insulator by quantum tunneling. Josephson predicted Cooper pair tunneling for the first time.

magnetic fluctuation. The reason for this is as follows:

The flow of electrons in HTSs is not by individual electrons, but by a large number of coupled electron pairs, or Cooper pairs. In LTSs, when one electron moves, it distorts the surrounding crystal lattice, which attracts another electron and makes a bond between the two electrons. Each Cooper pair is formed to minimize the distortion energy, and if thermal vibrational energy in the crystal lattice is less than this distortion energy, electricity by Cooper pairs flows without energy loss. This is allegedly to be the mechanism of superconductivity.*[168]

The mechanism of high temperature superconductivity is very similar to that of low temperature superconductivity. However, for high temperature superconductivity, there is no role of phonon and this role is replaced by spin-density waves (SDWs).*[169] Just as LTSs are strong phonon systems, HTSs are strong SDW systems near the magnetic transition from paramagnetism to antiferromagnetism. When a single

---

*[168] This is similar to the concept of normal electric conductivity in the new paradigm.

*[169] A SDW, together with charge-density wave (CDW), is a state of regular arrangement of a solid at low temperatures. This state is formed in metals with a high state density or in low dimensional anisotropic solids at the Fermi level. Other low temperature states forming in these materials are superconductivity, ferromagnetism and antiferromagnetism. The transition to alignment is due to the condensation energy, approximately $N(E_F)\Delta^2$., where $N(E_F)$ is the Fermi level and Δ is the energy gap for this state transition. SDWs are evident in spin waves, the excited states of ferromagnetic and antiferromagnetic materials. Essentially, SDWs and CDWs are accompanied by a hyperstructure in which the electron spin and charge density are modulated periodically. This superstructure is not translated according to the symmetry group that defines the location of ions.

electron moves in a HTS, the electron spin attracts SDWs around it. These waves again make a nearby electron belong to the spin generated by the first electron. A Cooper pairs is thus formed. At lower temperatures, more SDWs and Cooper pairs are formed and superconductivity appears. Since HTSs are magnetic materials due to Coulomb interactions, there is a strong repulsion between the electrons. Because of this repulsive force, no Cooper pairs are formed at the same lattice point, but are formed with electrons at the adjacent lattice points. This is the combination of so-called d-waves.

These are the pictures of high temperature superconductivity to date. However, the theories of low and high temperature superconductivity ignore the most important fact. Superconductivity is the electricity that flows only on (or near) the surface. In the interiors of a superconductor ε_P bonds have a enough high density to make a cross-linked network below T_c and thus the conductor is electromagnetically solidified. Therefore, the formation of magnetic fields in the interiors is suppressed, and electrical energy does not propagate in the form of wave. Our superconductivity by the new paradigm is pictured in the next section.

6.4. Unified theory of superconductivity

HTSs are mainly oxides, such as YBCO123 in Figure 97, and Cu is one of the main components. Why are they poorly conducting oxides, and why is Cu with a good conductivity one of the main components?

In the previous analysis of low temperature superconductivity, we showed that superconductivity requires secondary chemical bonds, namely ε_P bonds, not primary bonds, and for this purpose, available PeNs that do not participate in primary chemical bonds are required. The higher the strength of ε_P bonds (the higher the energy required to break the bond), the higher the critical temperature T_c and, in general, the higher T_c under higher pressures. Cu does not form any distinctive ε_P bonds, and, if any, the bonds does not make any networked structure leading to superconductivity.*[170] So how did Cu become a key element for high temperature superconductivity?

This can be interpreted that Cu makes chemical bonds with O (and thus the Cu atoms are attracted to the O atoms in YBCO), which distorts the location of Cu $3d$ PeNs, resulting in more available PeNs for forming distinct ε_P bonds. As temperature decreases (the lattice constant decreases), the density of ε_P bonds increases, which in turn suppresses the formation of magnetic field by electric currents. Hence the resistivity should increase rapidly as temperature decreases to T_c, as shown in Figure 86. Below T_c, an electromagnetic solidification by cross-linking of ε_P bonds may result in the infinite resistivity, but the resistivity disappears because the distortion of ε_P bonds by external electric fields is allowed only at or near the solid surface and magnetic fields are formed only in the vacuum.

---

*[170] The transportation of distorted and deformed ε_P bonds in a wave form is a current, and in the case of Cu the bonding energy of ε_P bond may be very low, or the density is too low to make a solid network even at very low temperatures.

Normal current and supercurrent have the same origin

The strength of ε_P bonds in HTSs is higher than that of LTSs, so it is more difficult to thermally disintegrate the ε_P bond network and thus T_c of HTSs should be higher. Though not considered in conventional theory, the suppression of magnetic field in superconductors is the key to understanding superconductivity. When electricity flows in a conductor, a magnetic field is accompanied, and according to Faraday's law, an induced electromotive force with the opposite sign is generated, and microscopic short circuits occur in the conductor. The generation of induced electromotive force means that the orientation of induced distortion of ε_P bonds is opposite to that of primary distortion of ε_P bonds by external electric fields, so that these two antiparallelly distorted states meets to transform electricity to thermal energy. To suppress the formation of the reverse distortion in a superconductor, magnetic fields should be formed only in the solid vacuum in contact with the conductor. However, HTSs with two-dimensional anisotropic structures such as YBCO have sufficient internal empty spaces, so magnetic fields induced by the primary distortion of ε_P bonds can also form in the internal solid vacuum. It does not generate resistant heat and magnetic fields appear to penetrate into the superconductor which will be explained in more detail later.

In summary, valence PeNs (not participating in primary bonds) responds to an external electric field, and secondary interatomic bonds are deformed to optimize the electric potential energy (Isotropic thermal vibration becomes anisotropic under the influence of the electric field). This local deformation induces an additional deformation called a magnetic field in the perpendicular plane to the orientation of the primary distortion of ε_P bonds. Electricity flows accompanied by the movement of this local deformation. Below T_c, any inductive deformation is not allowed in the solidified ε_P bond structure, so that electrical conduction in the conductor is blocked and the conductor becomes a perfect insulator. However, the network structure of ε_P bonds at the surface is flexible (different from the internal structure) and there is no electrical resistivity because magnetic fields are formed only in the solid vacuum in contact with the surface. In other words, superconductivity is the rearrangement of the ε_P bond structure at the surface following the external electric field and the induced deformation of the solid vacuum.*[171]

What YBCO superconductors tell

In terms of electrochemistry, an excited state (a distorted ε_P bond due to external electric fields) is a reduced state and the movement of the excited state is a kind of continuous redox reaction. Unlike LTSs, HTSs contain oxides, and oxygen will be involved in the redox reaction.

Let's consider the redox reaction that causes electrical conduction in YBCO, a representative HTS. In YBCO, if Y and B take O in stoichiometric ratio, three Cu atoms occupies 3.5 O atoms ($=½Cu_2O+CuO+CuO_2$) out of 7 in

---

*[171] Metaphorically speaking, normal electrical conductivity is the movement of a swimmer through liquid water, and superconductivity is the skating of a speed skater on solidified water, or ice. The player itself is a magnetic field.

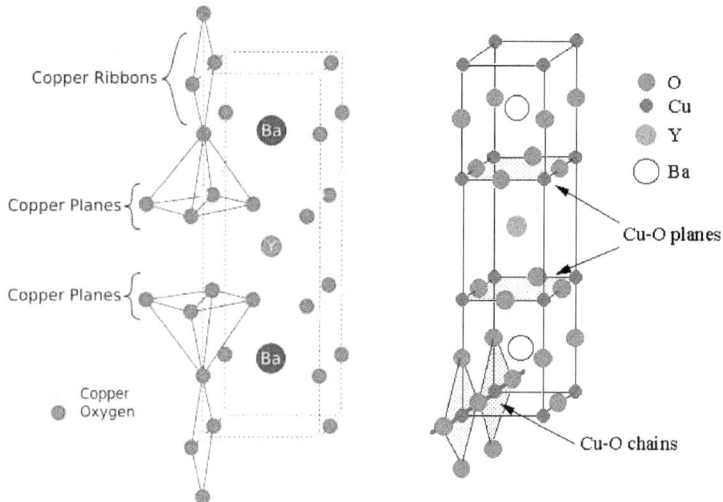

Figure 102. The arrangement of Cu and O atoms in the crystal structure of YBCO (image from wikipedia.org)

$YBa_2Cu_3O_7$. Of these three copper oxides, Cu_2O (cuprous oxide) is a hole type semiconductor. In other words, PeNs are insufficiently provided from O atoms. Cu_2O and CuO (cupric oxide) are stable oxides and have the crystal structures as shown in Figure 71. On the other hand, CuO_2 (copper peroxide) is an unstable oxide, which is easily decomposed into O and other copper oxides. In terms of oxidation state, Cu_2O is reduced one (excess PeNs for Cu), CuO_2 is oxidized one (lack of PeNs) and CuO is relatively oxidized or reduced with respect to the other oxides. When a Cu atom is reduced, it will change from Cu^{+4} (CuO_2) to Cu^{+2} (CuO) and further to Cu^{+1} (Cu_2O). Cu has the PeN configuration of [Ar] $3d^{10}$ $4s^1$, but in Cu_2O, the configuration should be $3d^{10}$ $4s^1$, in CuO $3d^9$ $4s^2$, and in

- 252 -

CuO_2 $3d^7$ $4s^2$ $4p^2$, respectively. The ½Cu_2O+CuO+CuO_2 structure would be any distorted one in the presence of Y or Ba, and thermal vibration would cause Cu to be repeatedly oxidized and reduced in dynamic equilibrium. When an electric field is applied, this vibration becomes anisotropic and energy propagates. For the flow of electricity, we may think of a steady reaction between Cu_2O and CuO_2 in the CuO_2 layer as

$$½Cu_2O+CuO_2 \Leftrightarrow 2CuO_{1.25} \Leftrightarrow CuO_2+½Cu_2O \quad --- (6.1).$$

YBCO has two CuO_2 layers in the unit cell separated by Y as shown in Figure 102. The CuO_2 layers cannot consist solely of pure CuO_2 because Y must also be present in a oxidized form. Therefore, it is natural for CuO_2 to coexist with other copper oxides as in Equation (6.1).[172] It is imagined that the two phases are repeatedly separated and combined to transfer energy. Eq. (6.1) means that when an electrical energy ΔE is input, the two phases change into a single high energy phase.[173] When this single phase separates into the two phases, it releases energy. When this process moves in one direction, electricity flows, in which the oxidation state of Cu and thus the location of the bond between Cu PeNs and O PeNs changes periodically.

---

[172] CuO will also be present but omitted to simplify the discussion.

[173] In Eq. (6.1), it is not known whether the two separated phases are energetically high or the single phase is high (thermodynamic calculations are required). If the single phase is energetically stable, copper oxides will exist as $CuO_{1.25}$ inside the superconductor below T_c, and in the opposite case the two phases will be separated.

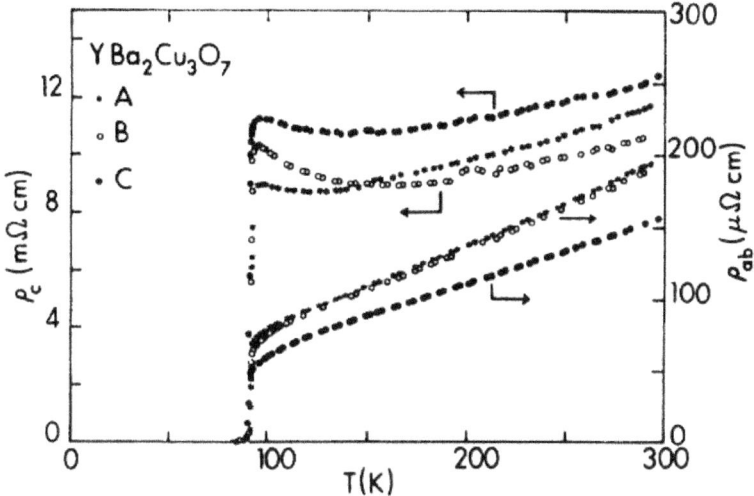

Figure 103. Resistivity of YBCO in the parallel (ρ_{ab}) and perpendicular directions (ρ_c) to the CuO_2 layer (three different specimens A, B and C, by S.J. Hagen, et al.)

As discussed in Section 5.2, for conductors, the lattice constant and the electrical resistivity decrease with a decrease in temperature. For insulators, on the other hand, the bond flexibility decreases and the resistivity increases. Referring to Figure 103, as temperature decreases, the resistivity in the plane parallel to the CuO_2 layer decreases, but the resistivity along the c-axis (perpendicular to the CuO_2 layer) does not decrease but increases as temperature decreases to T_c.[232] The bond flexibility in the Y an B oxides in this perpendicular direction decreases more rapidly, which can be interpreted as the redox reaction in Eq. (6.1) is difficult along the c-axis and thus the resistivity gets higher on approaching T_c.

The electrical resistivity of YBCO varies greatly depending on the crystallographic orientation as shown in Figure 103, but the critical temperature of superconductivity is not relevant to the crystallographic orientation. This isotropy means that the formation of magnetic field is suppressed in all directions in the superconductor below T_c. This also means that the distortion of ε_P bonds is only allowed at the solid surface independent of the crystallographic orientation. This isotropy in superconductivity is difficult to understand because it is thought that superconductivity does not appear or T_c will be lower along the c-axis. However, the critical magnetic field at which superconductivity disappears is 120 T in the direction perpendicular to the CuO_2 layer and 250 T in the parallel plane.[233] We can understand that the electromagnetically solidified network structure has an anisotropy like in the YBCO crystal structure and its bond strength is weak in one direction.

HTSs are II-type superconductors. Magnetic fields penetrate into the superconductor in the range between the primary and secondary critical magnetic field strength. How could we interpret this? As mentioned previously, the penetration of a magnetic field into the superconductor is interpreted as the formation of magnetic field in the solid vacuum contained in the interiors of the superconductor. This is a state in which superconductivity and magnetism coexist. If the strength of the external magnetic field exceeds that limit, the whole solid ε_P bond structure cannot sustain, and thus the electromagnetic solid become a liquid in which an induced secondary distortion of ε_P bonds can form. The superconductor becomes a normal conductor.

Below the secondary critical magnetic field, superconductivity is sustained for type II superconductors because the solid vacuum in the interiors of the superconductor can sufficiently accommodate the magnetic field. The large volume of the internal solid vacuum means a low mass density of the superconductor, which let us think that T_c would be higher for superconductors with lower mass densities. But this idea is not acceptable without objection because most superconductors have low critical temperatures at high pressures. Furthermore, as with Si in Figure 89, T_c does not increase simply with pressure. Taken together, in order to increase T_c, a superconductor should have a very anisotropic structure that secures enough volume of the internal solid vacuum that accommodates the magnetic field in one direction and that secures the number of effective PeNs for ε_P bonds in the other directions.

In this sense, if the c-axis lattice constant is large in superconductors with a cubic crystal structure, high critical temperatures are expected due to the high volume of the internal solid vacuum. An example is shown in Figure 104. It is seen that T_c increases with an increase in the c-axis lattice constant for InTe.[234] Most copper oxide superconductors except YBCO are tetragonal, and T_c is high when the c-axis lattice constant is large.[235] However, for YBCO with a high anisotropic crystallographic structure as shown in Figure 102, T_c decreases as the c-axis lattice constant increases. YBCO becomes a superconductor with the oxygen content of 7-x (0≤x≤0.65), and T_c is the highest as 95 K at x = 0.07. However, T_c decreases as the c-axis lattice constant increases with x, and when x is ~

0.6, the crystal structure changes to tetragonal and loses superconductivity. That is, for YBCO, T_c is sensitive to the ratio of the a- and b-axis (parallel to the CuO_2 layer) lattice constants.

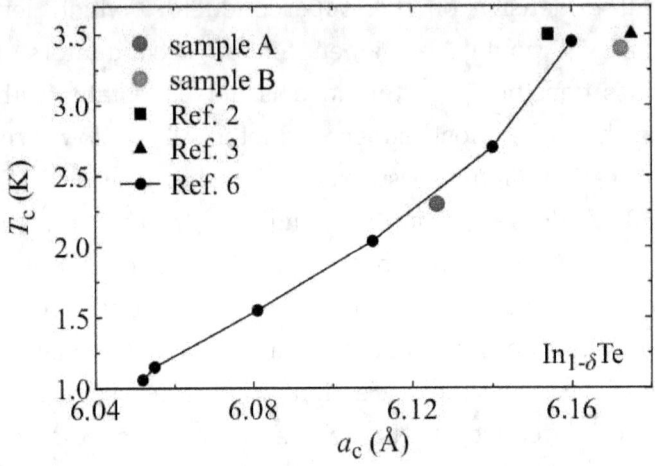

Figure 104. Critical temperature T_c as a function of the c-axis lattice constant (a_c) of cubic (M. Kriener et al.)

Mechanism of superconductivity in the new paradigm

The followings are summaries of the normal electric and superconducting phenomena understood in the regime of the new paradigm so far.
- First, the electrical conductivity is a viscoelastic behavior of interatomic ε_P bonds in solids (including some liquids). The deformation of ε_P bonds by an external electric field takes time for the rearrangement of the bonds in solids. As a result, thermal resistance is generated.

- Thermal resistance in a conductor is due to micro short-circuits generated by the electric field of opposite sign (reversely distorted ε_P bonds) induced by the magnetic field (the induced distortion of ε_P bonds) by an external electric field. In other words, thermal resistance is due to Faraday's law at the atomic level. If any thermal vibration has anisotropy in an external electric field, this anisotropic vibration is transferred to adjacent atoms, and the vibrations of the adjacent atoms are also anisotropic.
- As temperature decreases (the average interatomic spacing is reduced),[174] the solid becomes harder and more elastic in terms of the propagation of electric field (the propagation of vibration energy). For conductors, the pure elastic rearrangement of valence PeNs becomes dominant over the viscous rearrangement, and thus thermal resistivity is reduced. For semiconductors or insulators, the response to an external electric field becomes difficult (because there are no flexible secondary bonds). When temperature rises, the valence PeNs can be more easily rearranged and the electrical conductivity increases for semiconductors and insulators.
- For conductors, when temperature is extremely low, the formation of magnetic field (induced distortion by the primary distortion of ε_P bonds) is impossible due to the phase transformation (an electromagnetic solidification) at the critical density of ε_P bonds. Thus, the conductor becomes either a perfect insulator or a superconductor. The resistivity of some superconductors become zero after

---

[174] As in the cases where water becomes ice or iron (Fe) transforms from fcc to bcc, the average interatomic spacing can increase.

a sharp rise on going down to the critical temperature, T_c.
- Below T_c, ε_P bonds are distorted by external electric fields only near the surface (down to the London penetration depth) via interacting with the solid vacuum, the distortion of which is the magnetic field, so that electricity flows without resistivity only at the surface and its vicinity.
- At high pressures, T_c changes due to the decrease in the interatomic distance (the length of ε_P bond). When the bond strength or density of ε_P bonds are higher at higher pressures, we have higher T_c. But if the interatomic distance is too small, T_c rather becomes lower. There is an optimal pressure range to give the maximum ε_P bond strength and density, which depends on the crystal structure of superconductors and the number of valence PeNs of the constituting atoms.
- The crystal structure is influenced by pressure, and T_c is raised or lowered depending on the variation of the crystal structure. T_c increases due to the crystallographic change, in which the number of valence PeNs for ε_P bonds increases. On the contrary, T_c is lowered when some of PeNs participated in ε_P bonds are involved in primary chemical bonds upon crystallographic changes.
- Secondary interatomic ε_P bonds near the superconductor surface can be thermally excited, but the induced excitation (the magnetic field) forms only in the nearby solid vacuum, which is electromagnetically perfect elastomer, hence the electrical resistivity does not affected by this magnetic field developed in the solid vacuum.
- HTSs have highly two-dimensional anisotropic structures compared to LTSs. Primary chemical bonds are mainly covalent bonds with high bond energy, and can sustain the

anisotropic structures firmly. Due to this anisotropy, the electrical conductivity is also anisotropic and T_c is high due to the high density of valence PeNs capable of forming ε_P bonds and the high bonding strength.

- Transition metal oxides provide good conditions regarding to the number of PeNs involved in ε_P bonds and to the bond strength. Oxide-based superconductors have many available PeNs from the d-orbitals of the transition metal for ε_P bonds as a result of covalent bonding with oxygen. In addition, they have a crystal structure with a high anisotropy in which the variable oxidation state is maintained in the two-dimensional crystallographic plane. Since the distortion of ε_P bonds adjusting the external electric field and induced distortions by this primary distortion are easily made in the plane, the electrical conduction is efficient in the directions parallel to the plane while it is difficult in the perpendicular direction.

- Below T_c, ε_P bonds are three dimensionally networking, yielding an electromagnetically solidified structure and superconductivity occurs isotropically. Regardless of whether it is a LTS or a HTS, magnetic fields are formed only in the solid vacuum, which is a perfect electromagnetic elastomer, and electricity flows only through the surface layers. HTSs are high in the density of ε_P bonds even at high temperatures because the electromagnetically solidified structure is well preserved in the two dimensional crystal structures.

- T_c is sensitive to the specific crystal structure formed by primary bonds. When a HTS has a highly anisotropic crystal structure with a bulky internal free space (the solid vacuum), the critical temperature and the critical magnetic field are high. This is because the internal solid vacuum can accommodate the magnetic field induced by the distortion of ε_P bonds. When the crystal is tetragonal or cubic, T_c increases as the lattice constant along the c-axis increases, but when the lattice constants along the a- and b-axis change, the interatomic distance changes, affecting the number of available PeNs for ε_P bonds. The lattice constant may vary depending on the content of the constituting elements, which in turn influences T_c. As shown in Figure 105, the lattice constant caused by the change of nitrogen content is closely related to T_c for V_3PN_x or V_3AsN_x.[236]

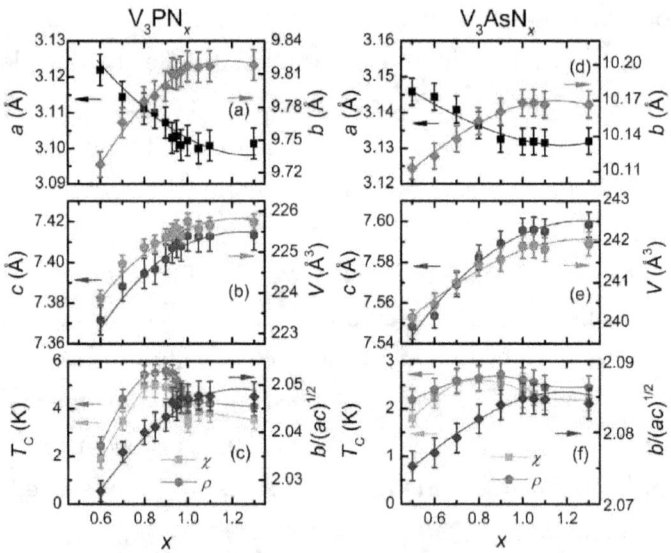

Figure 105. Lattice constant and T_c as a function of nitrogen content (x) (B. Wang, K. Ohgushi).

VII. For room temperature superconductors

In our previous book, "Origin of Gravity and New Cosmos (ISBN 9781713042020, 12.2019)", in the regime of the new vacuum paradigm that the vacuum is made of a very dense matter and devoid of energy, we addressed the origin of gravity and reinterpreted cosmological phenomena. When energy is added to the solid vacuum, it is deformed and distorted as much as the amount of energy. Resultantly, stresses develop not only in the solid vacuum and but in ordinary matter, and the variation of the stress is the origin of all physical phenomena in the new vacuum paradigm. The solid vacuum does not convert to any other things, but the energy stored in it varies upon changing the center of vibration in the vacuum lattice, and the transfer of matter or energy is nothing but the transfer of stored energy in the solid vacuum. This book attempted to interpret the electromagnetic behaviors of materials in terms of their interaction with the solid vacuum. We proposed a new atomic model, and based on this model, we newly analyzed the characteristics of chemical bonding, including hydrogen bonding, to grasp electrical and magnetic phenomena based on the analysis, and to present a new paradigm for electric conductivity and superconductivity.

New atomic model and periodicity

When free electrons are captured by protons, they form hydrogen atoms. Hydrogen atoms turn into neutrons by fusion under high pressures and temperatures inside stars, producing helium and other heavy elements. Since neutrons formed by nuclear fusion are depleted of the energy of hydrogen atoms, helium containing two neutrons is a very stable element. The conventional atomic model for hydrogen and helium is that the electrons orbit the atomic nucleus composed of neutrons and protons. The exact position of the electron is impossible to specify, and only the probability calculation using quantum mechanical electron wave functions is allowed. For heavy elements, even mathematical interpretation is impossible, and it is hard to imagine that nucleons are gathered in the center and many electrons roam around them. In our book, under the hypothesis that the neutrons generated via fusion combine hydrogen atoms to form heavy atoms, a neutron-hydrogen atom[*175)] pair makes a building block for heavy atoms, and a new atomic model is proposed in which these building blocks are stacked in a spherical form to yield various kinds of atom. In this model, the atomic number is the number of the building blocks, namely PeNs (PeN = plecton + neutron). Atoms composed of PeNs become spherical to accommodate with the surrounding solid vacuum. In the periodic table of elements, the number of atoms that can be staked in one period increases exactly in proportion to the square of radius. Helium, neon, argon

---

[*175)] A pair of electron-proton in an atom are called plecton in this book.

atoms are in spherical forms, and their stress fields in the surrounding solid vacuum are symmetric, For other non-spherical atoms, stresses are developed asymmetrically in the solid vacuum. To mitigated these asymmetrical stresses, the atoms chemically bond together. The heaviest element discovered or synthesized so far is organoneson (Og, atomic number 118) in the 7th period, but higher atomic number elements can be found, and neutron stars may be very heavy elements.

Interatomic force (chemical bonds)

Heavier elements than hydrogen were born by piling up PeNs, the basic building blocks, to form energetically stable spheres, like neon, argon, krypton, etc. These elements do not react chemically with other elements. However, most elements are not spherical and thus unstable, distorting the surrounding solid vacuum asymmetrically. Unstable free atoms combine with other atoms to form molecules, metals, etc. to be stabilized together. Valence PeNs of one atom interact with valence PeNs of neighboring atoms to form chemical bonds. The combination of two valence PeNs of two adjacent atoms yields a helium-like structure. The type of chemical bond depends on this specific helium-like structure resulting in the difference in the bond strength.

It is understood that the PeNs participating in metallic bonding are not settled in one location and constantly move around the atomic surface because the atom is not sufficiently supplied with valence PeNs from adjacent atoms of the same kind. Therefore, the bond strength is relatively weak compared to other types of chemical bond, although the bond strength increases with the number of available PeNs. When a metallic element combines with a non-metallic one, an ionic bond is formed, This bond is very localized because the valence PeN of the metallic element is well fit to the location where a PeN is deficient for the non-metallic element. So they chemically bond, but are weak as a solid.

If one atom shares one PeN with another atom's PeN, and the location is fixed, the bonding is very strong. It is a covalent bond. If this bond constitutes a diamond structure with four pairs of PeNs, its mechanical strength is very high. Bonding between hydrogen and other elements makes also a covalent bond. In this case, the resultant combination is ^{3}He, a helium isotope that lacks a neutron, so molecules covalently bound to hydrogen form an additional type of bonding known as hydrogen bonding. This is a kind of secondary chemical bonding. The van der Waals force can be understood also as a type of weak secondary bonds (weaker than hydrogen bonds) formed between interatomic PeNs that do not or partially participate in primary bonds, like ionic or covalent bonds. Weak primary metallic bonds or secondary bonds are sensitive to the environmental conditions (temperature, pressure, electromagnetic field, etc). The interaction with the environment determines the physical properties of solids.

Electromagnetic behaviors of solids

Electrical and magnetic properties of materials are the results of their interaction with the surrounding solid vacuum. The electrical conductivity of materials is the movement of distorted secondary interatomic PeN bonds (we called them ε_P bonds) in an external electric field. The distorted bonding is a high energy state, and moves through a thermally vibrating medium in the direction of the external electric field.

Thermal random vibrations of ε_P bonds in a solid have no preferred orientation, and thus the flow of thermal energy is isotropic. The state of ε_P bonds (e.g. the average bond length) is in thermal equilibrium. The bonds are constantly stressed and relaxed in a random manner. When an external electric field is applied, ε_P bonds are distorted and the random vibration becomes anisotropic, restricted by the external electric field. When the anisotropic vibration of ε_P bonds is localized at the atomic level, a new equilibrium is set up. However, when they are moving through the material along the external electric field, electricity flows. The distortion of a ε_P bond is structurally hampered by primary and other secondary bonds and the distortion is made in a viscoelastic manner.

Changing the relative position of a valence PeN of the constituting atom in a solid also affects those of neighboring atoms. When a solid is tensioned in one direction due to this change, the solid is contracted in the perpendicular plane, and vice versa. The response in the plane perpendicular to an electric field is the magnetic field. Electricity and magnetism are inseparable and are integrated into electromagnetism. Light is a wave that propagates as the distortion of the solid vacuum changes periodically, and a change in one direction (an electric field) is accompanied by a change in the perpendicular plane (a magnetic field). This wave propagates in the same way in the solid vacuum or matter, but within matter, valence PeNs play a role in the propagation, so its speed is different from that in the vacuum. When electricity flows, a magnetic field is inevitably generated due to the change of location of the valence PeNs on the atomic surface. The flow of electricity is the flow of potential energy stored in ε_P bonds which are repeatedly stressed and relaxed. In this process, if the deformation is purely elastic, there will be no electrical resistance, but because materials do not deform purely elastic (from an electromagnetic point of view), part of the stored energy is converted into heat by the inelastic behaviors of materials. It is the electrical resistivity.

ε_P bonds in an electrical conductor are easily deformed or distorted in response to an external electric field, and the distortion propagates very quickly. Insulators, on the other hand, are very stiff in this regard. Thermal resistance is generated in the conductor because time is required for the inelastic rearrangement of ε_P bonds when an external electric field is applied.

As temperature increases, the lattice constant of a conductor increases, leading to a decrease in the moving speed of the distortion energy of ε_P bonds, so that the resistivity increases, while insulators or semiconductors are facilitated in deforming the interatomic bonds by thermal vibration, making it easier to react to an electric field, and thus the resistivity decreases with an increase in temperature. In this sense, all solids decrease in resistivity with increasing temperatures at cryogenic temperatures and then increase above certain temperatures. If the temperature of this transition point is very low, it is a conductor, if it is a medium, it is a semiconductor, and if it is very high, it is an insulator.

Superconducting phenomena

Superconductivity is the flow of electricity without resistance below a certain temperature, the critical temperature. As temperature drops, the density of effective ε_p bonds in a conductor increases and the whole bond structure become harder. If the density is above the threshold value, the conductor is electromagnetically solidified, and thermal vibrations are not enough to break up this solidified structure. As a liquid becomes a solid, it becomes an electromagnetically hard solid. Just as the solid-liquid phase transformation is the breakdown of the structure formed by interatomic primary bonds, the phase transformation that loses superconductivity above the critical temperature is the breakdown of the network structure based on ε_p bonds.

If an external electric field is applied above the critical temperature, the valence PeNs of ε_p bonds are relocated in a viscoelastic manner. A viscoelastic behavior means that it takes time to set up a new equilibrium for the relocation of the PeNs in ε_p bonds. In order for electricity to flow, additional distortions of ε_p bonds in the plane perpendicular to the applied external electric field are inevitable. These induced distortions show up as a magnetic field. When a magnetic field is induced in a conductor due to the flow of electricity, an electric field with the opposite sign according to Faraday's law is generated (the orientation of distortion is reverse to the primary distortion of ε_P bonds by the external electric field), resulting in micro short circuits and yielding resistant heat. However, below the critical temperature, ε_P bonds are cross-liked and the whole structure is electromagnetically solidified. Therefore the valence PeNs cannot rearrange in the interiors of the conductor, if the strength of external electric or magnetic field is not big enough, and electricity cannot flow in the interiors of a superconductor.

However, the formation of magnetic field in the solid vacuum near the surface of a superconductor is allowed, electricity can flow through the surface layers of the superconductor without resistance, as there are no internal magnetic field and thus no internal electric field induced by Faraday's law. As solids contains some free space in the interiors, magnetic fields appear to penetrate to some extent into the superconductor. This is the London penetration depth.

Basically, low temperature superconductivity and high temperature superconductivity are the same phenomenon. High temperature superconductors have favorable crystal structures with regard to the critical temperature, such as copper oxide YBCO123. Electricity flows preferentially through the CuO_2 layers. When ε_P bonds in the CuO_2 layer are anisotropically distorted in response to an external electric field, a magnetic field, an induced distortion of ε_P bonds, is also formed and constrained on the plane parallel to the CuO_2 layer. In other words, electric and magnetic fields develop mainly on the same plane. This is because the rearrangement of the PeNs in ε_P bonds in the orientation perpendicular to this plane is comparatively difficult due to the presence of other oxide layers in the YBCO stack. Therefore, the electrical conductivity depends on the crystallographic orientation.

Since most of ε_P bonds interacting with the electric field are constrained two-dimensionally, the bond density is higher in the layers of HTSs, while for LTSs the bond density has no such an anisotropy. For YBCO, electrical conduction occurs when the oxidation state of copper changes in the oxygen-deficient CuO_2 layer. So the layer is a composite of the three copper oxides, Cu_2O, CuO and CuO_2. Depending on the fraction of oxygen depleted, the ratio between these three oxides at equilibrium will vary, and the bonding structure will react differently to temperature and external electromagnetic fields. Above the critical temperature, the oxidation state of copper is in dynamic equilibrium. The fraction of each copper oxide varies ceaselessly, but the average fraction is constant, which may be regarded as a single phase of copper oxide. When temperature is lowered, the phases are separated and fixed and the three compounds exist independently, so the flow of electricity in the interiors of the superconductor is blocked. However, since the crystal structure at the surface is unstable compared to the bulk structure, the oxidation state of copper can respond to the external electric field and move without resistance.

Below the critical temperature at which ε_P bonds make a cross-linked structure to yield an electromagnetic solid, the conductor becomes a perfect insulator or superconductor. What makes a conductor be a superconductor? The ε_P bonds in the surface layer are flexible two-dimensionally, so that they are distorted differently from those in the interiors, and the magnetic field induced by these distortions should be formed only in the solid vacuum. Some magnetic fields can penetrate into superconductors because they contain in their structure a sufficient volume of the solid vacuum to accommodate the internally developed magnetic field.

For room temperature superconductors

Superconductivity is electrical conductivity without resistivity in which electricity flows only through the surface layers of the conductor at low temperatures (below the critical temperature). No internal magnetic fields form, because an electromagnetic solidification occurs via networking of ε_P bonds at the critical temperature. The superconductor reacts to external electric fields below the critical temperature, by distorting individual ε_P bonds only in the surface layers (with the formation of magnetic field in the solid vacuum), and this distortion propagates in the direction of the external electric field. Electricity flows without resistance because magnetic fields are induced only in the solid vacuum, an electromagnetically perfect elastomer. Superconductivity is an electrical conduction that occurs at the interface between the conductor and the solid vacuum below the critical temperature. Hence, the dream materials, room temperature superconductors, must meet the following conditions.

A. The energy to break the ε_P bond structure must be high, which requires a large number of available valence PeNs for ε_P bonds, in which the bonding strength should be high.

B. Enough amounts of the solid vacuum should be included in the crystal structure. The induced distortion of ε_p bonds (the induced magnetic field) in response to the primary distortion of ε_p bonds by an electric field should be confined in the internal solid vacuum.

C. In order for the distortion of ε_p bonds to propagate easily, superconductors must be crystallographically unstable (in whole or in part).

Room temperature superconductors satisfying the above conditions should have highly anisotropic crystal structures.

- To satisfy the condition "A", the crystal must contain transition metal elements with a large number of available PeNs that do not participate in primary interatomic bonds.
- To satisfy the condition "B", the lattice constant in one direction (e.g. the c-axis) must be as large as possible compared to the other lattice constants. In this case, the number of available PeNs for ε_p bonds should not be reduced.
- To satisfy the condition "C", the number of the crystal structures formed based on primary bonds should be more than two and the structures should be variable according to the external conditions (temperature, pressure, electromagnetic field). The crystal structure should be unstable and distorted one in itself.

In order to satisfy these conditions, the specific crystal structure should be as follows.

- Possibly thinner insulating and conducting layers are

stacked (available PeNs are concentrated in the layer of a kind, facilitating electromagnetic solidification by the networking of ε_p bonds) and the primary bonds in the insulating layer should be also flexible to some extent.

-The insulating layer has a planar structure with covalent bonds and the crystal structure of the upper and lower insulating layers is different, which makes the crystal structure of the conductive layer unstable (the conductive layer is in a distorted state even in the absence of external electromagnetic fields).

- The conducting layer should be made of transition metal oxides with as many available PeNs as possible for ε_p bonds in the crystal structure formed based on primary bonds (the oxidation state of transition metals must be easily variable).

-There must be a lot of empty space inside the superconductor, so the inner surface area should be wide. (The lattice constant of the c-axis perpendicular to the conductive layer should be large, but the change in the lattice constants of the a and b-axes should not lower the number of available PeNs for ε_p bonds.)

Reference

1 Diamond (C) - Properties, Applications, www.azom.com/properties.aspx?ArticleID=262.

2 C. Broggini (2003). Physics in Collision, Proceedings of the XXIII International Conference: Nuclear Processes at Solar Energy. XXIII Physics in Collisions Conference. Zeuthen, Germany. p. 21.

3 A.H. Guth (1998). The Inflationary Universe: The Quest for a New Theory of Cosmic Origins. Basic Books. p. 186. ISBN 978-0201328400. OCLC 35701222.

4 D. Cirigliano, H.J. de Vega, N.G. Sanchez (2018). "Clarifying inflation models: The precise inflationary potential from effective field theory and the WMAP data". Physical Review D. 71 (10): 77-115.

5 S. Weinberg (2002). Foundations, The Quantum Theory of Fields, I, Cambridge University Press, ISBN 0-521-55001-7.

6 The University of North Carolina at Chapel Hill. "Nuclear Chemistry". Retrieved 2012-06-14.

7 G.A. Miller (2007). "Charge densities of the neutron and proton", Phys. Rev. Lett. 99, 112001.

8 H.M. Leicester (1971). The Historical Background of Chemistry. Courier Dover. pp. 221-222. ISBN 0-486-61053-5.

9 ibid 8.

10 Frank Wilczek: "Happy Birthday, Electron" Scientific American, June 2012.

11 J.J. Thomson (1897). "Cathode Rays". Philosophical Magazine. 44 (269): 293-316.

12 W.G. Myers (1976). "Becquerel's Discovery of Radioactivity in 1896". Journal of Nuclear Medicine. 17 (7): 579-582.

13 L.J. Curtis (2003). Atomic Structure and Lifetimes: A Conceptual Approach. Cambridge University Press. p. 74. ISBN 0-521-53635-9.

14 1. M. Agostini, et al. (Borexino Coll.) (2015). "Test of Electric Charge Conservation with Borexino". Physical Review Letters. 115 (23): 231802.
2. J. Beringer (Particle Data Group), et al. (2012). "Review of Particle Physics: [electron properties]". Physical Review D. 86 (1): 010001.
3. H.O. Back, et al. (2002). "Search for electron decay mode e → γ + ν with prototype of Borexino detector". Physics Letters B. 525: 29-40.

15 W. Greiner (2000). Relativistic Quantum Mechanics. Wave Equations (3rd ed.). Springer Verlag. ISBN 3-5406-74578.

16 P.A.M. Dirac (1930). "A Theory of Electrons and Protons". Proc. R. Soc. Lond. A. Royal Society Publishing. 126 (801): 360-365.

17 C.D. Anderson (1933). "The Positive Electron". Physical Review. 43 (6): 491-494. doi:10.1103/PhysRev.43.491.

18 E. Bland (2008). "Laser technique produces bevy of antimatter". MSNBC. Retrieved 6 April 2016. "The LLNL scientists created the positrons by shooting the lab's high-powered Titan laser onto a one-millimeter-thick piece of gold."

19 J. Palmer (2011). "Antimatter caught streaming from thunderstorms on Earth". BBC News. Archived from the original on 12 January 2011. Retrieved 11 January 2011.

20 D.W. Engelkemeir, K.F. Flynn, L.E. Glendenin (1962). "Positron Emission in the Decay of K40". Physical Review. 126 (5): 1818. doi:10.1103/PhysRev.126.1818.

21 H. Nishino, et al. (2009). "Search for Proton Decay via p→e+π^0 and p→μ+π^0 in a Large Water Cherenkov Detector". Physical Review Letters. 102 (14): 141801.

22 1. W. Weise, A.M. Green (1984). Quarks and Nuclei. World Scientific. pp. 65-66.
2. P. Ball (2008). "Nuclear masses calculated from scratch".

Nature. Retrieved Aug 27, 2014.
3. M. Reynolds (2009). "Calculating the Mass of a Proton". CNRS international magazine. CNRS (13). ISSN 2270-5317. Retrieved Aug 27, 2014.

23 A. Antognini, et al. (2013). "Proton Structure from the Measurement of 2S-2P Transition Frequencies of Muonic Hydrogen". Science. 339 (6118): 417-20.

24 1. W. Bothe, H. Becker (1930). "Künstliche Erregung von Kern-γ-Strahlen" [Artificial excitation of nuclear γ-radiation]. Zeitschrift für Physik. 66 (5-6): 289-306.
2. H. Becker, W. Bothe (1932). "Die in Bor und Beryllium erregten γ-Strahlen" [Γ-rays excited in boron and beryllium]. Zeitschrift für Physik. 76 (7-8): 421-438.

25 J. Chadwick (1933). "Bakerian Lecture. The Neutron". Proceedings of the Royal Society A: Mathematical, Physical and Engineering Sciences. 142 (846): 1-25.

26 P.J. Mohr, B.N. Taylor, D.B. Newell (2014). "The 2014 CODATA Recommended Values of the Fundamental Physical Constants" (Web Version 7.0). The database was developed by J. Baker, M. Douma, and S. Kotochigova. (2014). National Institute of Standards and Technology, Gaithersburg, Maryland 20899.

27 B. Povh, K. Rith, C. Scholz, F. Zetsche (2002). Particles and Nuclei: An Introduction to the Physical Concepts. Berlin: Springer-Verlag. p. 73.

28 J.-L. Basdevant, J. Rich, M. Spiro (2005). Fundamentals in Nuclear Physics. Springer. p. 155.

29 G.L. Greene, et al. (1986). "New determination of the deuteron binding energy and the neutron mass". Phys. Rev. Lett. 56: 819-822.

30 Basic Ideas and Concepts in Nuclear Physics: An Introductory Approach, Third Edition; K. Heyde Taylor & Francis 2004.

31 Y. Gell, D.B. Lichtenberg (1969). "Quark model and the magnetic moments of proton and neutron". Il Nuovo Cimento A. Series 10. 61: 27-40.

32 L.W. Alvarez, F. Bloch (1940). "A quantitative determination of the neutron magnetic moment in absolute nuclear magnetons". Physical Review. 57: 111-122.

33 D.H. Perkins (1982). Introduction to High Energy Physics. Addison Wesley, Reading, Massachusetts. pp. 201-202. ISBN 978-0-201-05757-7.

34 G.A. Miller, ibid 7.

35 "Pear-shaped particles probe big-bang mystery" (Press release). University of Sussex. 20 February 2006. Retrieved 2009-12-14.

36 E. Rutherford (1920). "Nuclear Constitution of Atoms". Proceedings of the Royal Society A. 97 (686): 374-400.

37 O. Klein (1929). "Die Reflexion von Elektronen an einem Potentialsprung nach der relativistischen Dynamik von Dirac". Zeitschrift für Physik. 53 (3-4): 157-165.

38 Daya Bay Collaboration (2016-02-12). "Measurement of the Reactor Antineutrino Flux and Spectrum at Daya Bay". Physical Review Letters. 116 (6): 061801.

39 S. Berryman, "Ancient Atomism", Stanford Encyclopedia of Philosophy (Fall 2008 Edition), Edward N. Zalta (ed.).

40 J.J. Thomson (1897). "Cathode rays", Philosophical Magazine. 44 (269): 293-316.

41 E.T. Whittaker (1951). A history of the theories of aether and electricity. Vol 1, Nelson, London.

42 J.J. Thomson (1904). "On the Structure of the Atom: an Investigation of the Stability and Periods of Oscillation of a number of Corpuscles arranged at equal intervals around the Circumference of a Circle; with Application of the Results to the Theory of Atomic Structure". Philosophical Magazine. 7 (39): 237.

43 H. Geiger (1910). "The Scattering of the α-Particles by Matter". Proceedings of the Royal Society. A 83: 492-504.

44 E. Rutherford (1911). "The Scattering of α and β Particles by Matter and the Structure of the Atom" (PDF). Philosophical

Magazine. 21 (4): 669.

45 N. Bohr (1913). "On the constitution of atoms and molecules" (PDF). Philosophical Magazine. 26 (153): 476-502.

46 E. Schrodinger (1926). "Quantisation as an Eigenvalue Problem". Annalen der Physik. 81 (18): 109-139.

47 Biographies of Great Man. " Born: Founder of Lattice Dynamics". Jan. 1, 2010. liferkurry.blogspot.com.

48 W. Heisenberg (1927). "Über den anschaulichen Inhalt der quantentheoretischen Kinematik und Mechanik". Zeitschrift für Physik (in German). 43 (3-4): 172-198.

49 D. Palmer (13 September 1997). "Hydrogen in the Universe". NASA. Retrieved 5 February 2008.

50 D.P. Stern (13 February 2005). "Wave Mechanics". NASA Goddard Space Flight Center. Retrieved 16 April 2008.

51 1. P. Atkins, J. de Paula, Atkins' Physical Chemistry, 8th edition (W.H.Freeman 2006), p. 451-2 ISBN 0-7167-8759-8.
2. M.J. Matthews, G. Petitpas, S.M. Aceves (2011). "A study of spin isomer conversion kinetics in supercritical fluid hydrogen for cryogenic fuel storage technologies". Appl. Phys. Lett. 99 (8): 081906. doi:10.1063/1.3628453.

52 V.I. Tikhonov, A.A. Volkov (2002). "Separation of Water into Its Ortho and Para Isomers". Science. 296 (5577): 2363. PMID 12089435.

53 J. Hritz (March 2006). "CH. 6 - Hydrogen" . NASA Glenn Research Center Glenn Safety Manual, Document GRC-MQSA.001. NASA. Retrieved 5 February 2008.

54 G.L. Miessler, D.A. Tarr (2003). Inorganic Chemistry (3rd ed.). Prentice Hall. ISBN 0-13-035471-6.

55 M. Salaris, S. Cassisi (2005), Evolution of stars and stellar populations, John Wiley and Sons, pp. 119-121, ISBN 0-470-09220-3.

56 I. Ahmad (1971). "The Proton type-nuclear fission reaction", The Nucleus, 1: 42, 59,

57 K.S. Krane, Introductory Nuclear Physics, Wiley, 1987, p. 537.

58 C.E. Chase, G.O. Zimmerman (1973). "Measurements of P-V-T and Critical Indices of He3". Journal of Low Temperature Physics. 11 (5-6): 551. doi:10.1007/BF00654447.

59 C.S. Nash (2005). "Atomic and Molecular Properties of Elements 112, 114, and 118". Journal of Physical Chemistry A. 109 (15): 3493-3500.

60 P. Ehrhart (1991). "Properties and interactions of atomic defects in metals and alloys", chapter 2, p. 88 in Landolt-Börnstein, New Series III, Vol. 25, Springer, Berlin.

61 University of Virginia, MSE 6020: Defects and Microstructure in Materials, Leonid Zhigilei.

62 J.C.M. Li, in Electron Microscopy and Strength of Crystals, eds. G. Thomas and J. Wschbum (New York: Interscience Publishers, 1963.

63 http://chemwiki.ucdavis.edu/Physical_Chemistry/Quantum_Mehns/Quantum_Theory/Trapped_Particles/Atoms/Quantum_Numbers

64 P. Halpern (2017-11-21). "Spin: The Quantum Property That Should Have Been Impossible". Forbes. Starts With A Bang.

65 Chemical Bonds. chemguide.co.uk

66 Metallic bonding. chemguide.co.uk

67 "Elastic Properties and Young Modulus for some Materials". The Engineering ToolBox.

68 periodictable.com/Elements/011/data.html

69 periodictable.com/Elements/011/data.html

70 periodictable.com/Elements/012/data.html

71 "Elastic Properties and Young Modulus for some Materials". The Engineering ToolBox.

72 periodictable.com/Elements/013/data.html

73 Arther Beiser, Concept of modern physics. 6th ed. McGraw-Hill (2003) p. 342

74 www.korth.de/index.php/162/items/24.html

75 "Chemical Bonds". Hyperphysics.phy-astr.gsu.edu.

76 www.chm.bris.ac.uk/motm/diamond/diamprop.htm

77 www.thoughtco.com/table-of-electrical-resistivity-conductivity-608499

78 www.azom.com/properties.aspx?ArticleID=516

79 G. Eranna (2014). Crystal Growth and Evaluation of Silicon for VLSI and ULSI. CRC Press. p. 7. ISBN 978-1-4822-3281-3.

80 M.A. Hopcroft, W.D. Nix, T.W. Kenny (2010). "What is the Young's Modulus of Silicon?". J. of Microelectromechanical Systems. 19 (2): 229. doi:10.1109/JMEMS.2009.2039697.

81 T. Steiner (2002). "The Hydrogen Bond in the Solid State". Angew. Chem. Int. Ed. 41: 48-76.

82 1. J.W. Larson, T.B. McMahon (1984). "Gas-phase bihalide and pseudobihalide ions. An ion cyclotron resonance determination of hydrogen bond energies in XHY- species (X, Y = F, Cl, Br, CN)". Inorganic Chemistry. 23 (14): 2029-2033. doi:10.1021/ic00182a010.
2. J. Emsley (1980). "Very Strong Hydrogen Bonds". Chemical Society Reviews. 9 (1): 91-124. doi:10.1039/cs9800900091.

83 P. Hobza, Z. Havlas (2000). "Blue-Shifting Hydrogen Bonds". Chem. Rev. 100 (11): 4253-4264. doi:10.1021/cr990050q.

84 H. Lodish, A. Berk, S.L. Zipursky, P. Matsudaira, D. Baltimore, J. Darnell (2000). "Molecular Cell Biology, 4th edition", New York: W. H. Freeman; ISBN-10: 0-7167-3136-3.

85 E. Pastorczak, C. Corminboeuf (2017). "Perspective: Found in translation: Quantum chemical tools for grasping non-covalent interactions", The Journal of Chemical Physics. AIP Publishing LLC, 146(12), p. 120901. doi: 10.1063/1.4978951.

86 F. Cordier, M. Rogowski, S. Grzesiek, A. Bax (1999).

"Observation of through-hydrogen-bond $^{2h}J_{HC'}$ in a perdeuterated protein". J Magn Reson. 140 (2): 510-2. doi:10.1006/jmre.1999.1899.

87 W.L. Jorgensen, J.D. Madura (1985). "Temperature and size dependence for Monte Carlo simulations of TIP4P water". Mol. Phys. 56 (6): 1381. doi:10.1080/00268978500103111.

88 J. Zielkiewicz (2005). "Structural properties of water: Comparison of the SPC, SPCE, TIP4P, and TIP5P models of water". J. Chem. Phys. 123 (10): 104501. doi:10.1063/1.2018637.

89 P.F. Dillon (2012). Biophysics: A Physiological Approach. Cambridge University Press. p. 37. ISBN 978-1-139-50462-1.

90 M.L. Cowan, B.D. Bruner, N. Huse, et al. (2005). "Ultrafast memory loss and energy redistribution in the hydrogen bond network of liquid H_2O". Nature. 434 (7030): 199-202. doi:10.1038/nature03383.

91 J. Luo, A.H. Jensen, N.R. Brooks, et al. (2015). "1,2,4-Triazolium perfluorobutanesulfonate as an archetypal pure protic organic ionic plastic crystal electrolyte for all-solid-state fuel cells". Energy & Environmental Science. 8 (4): 1276. doi:10.1039/C4EE02280G.

92 "Law-breaking liquid defies the rules", Archived 2011-04-29 at the Wayback Machine. Physicsworld.com (September 24, 2004).

93 W. Jencks, W.P. Jencks (1986). "Hydrogen Bonding between Solutes in Aqueous Solution". J. Am. Chem. Soc. 108 (14): 4196. doi:10.1021/ja00274a058.

94 M. Hellgren, C. Kaiser, S. de Haij, A. Norberg, J.O. Hoog (2007). "A hydrogen-bonding network in mammalian sorbitol dehydrogenase stabilizes the tetrameric state and is essential for the catalytic power". Cellular and Molecular Life Sciences. 64 (23): 3129-38. doi:10.1007/s00018-007-7318-1.

95 D. Laage, J.T. Hynes (2006). "A Molecular Jump Mechanism for Water Reorientation". Science. 311 (5762): 832-5. doi:10.1126/science.1122154.

96 O. Markovitch, N. Agmon (2008). "The Distribution of Acceptor

and Donor Hydrogen-Bonds in Bulk Liquid Water". Molecular Physics. 106 (2): 485. doi:10.1080/00268970701877921.

97 T. Steiner (2002). "The Hydrogen Bond in the Solid State". Angewandte Chemie International Edition. 41: 48. doi:10.1002/1521-3773(20020104)41:1<48::AID-ANIE48>3.0.CO;2-U.

98 G. Gilli, P. Gilli (2000). "Towards an unified hydrogen-bond theory". Journal of Molecular Structure. 552 (1-3): 1-15. doi:10.1016/S0022-2860(00)00454-3.

99 B. Schiøtt, B.B. Iversen, G.K. Madsen, F.K. Larsen, T.C. Bruice (1998). "On the electronic nature of low-barrier hydrogen bonds in enzymatic reactions". Proc. Natl. Acad. Sci. U.S.A. 95 (22): 12799-802. doi:10.1073/pnas.95.22.12799.

100 R.H. Garrett, C.M. Grisham (2016). Biochemistry (6th ed.). University of Virginia. pp. 12-13.

101 A.A. Abrikosov, L.P. Gorkov, I.E. Dzyaloshinsky (1963-1975). Methods of Quantum Field Theory in Statistical Physics. Dover Publications. ISBN 978-0-486-63228-5. Chapter 6 Electromagnetic Radiation in an Absorbing Medium.

102 J.E. Lennard-Jones (1924), "On the Determination of Molecular Fields", Proc. R. Soc. Lond. A, 106 (738): 463-477, doi:10.1098/rspa.1924.0082.

103 W.H. Keesom (1915). "The second virial coefficient for rigid spherical molecules whose mutual attraction is equivalent to that of a quadruplet placed at its center". Proceedings of the Royal Netherlands Academy of Arts and Sciences. 18: 636-646.

104 F.L. Leite, C.C. Bueno, A.L. Da Róz, E.C. Ziemath, O.N. Oliveira (2012). "Theoretical Models for Surface Forces and Adhesion and Their Measurement Using Atomic Force Microscopy". International Journal of Molecular Sciences. 13 (12): 12773-856. doi:10.3390/ijms131012773.

105 1. P.H. Blustin (1978). "A Floating Gaussian Orbital calculation on argon hydrochloride (Ar·HCl)". Theoretica Chimica Acta. 47 (3): 249-257. doi:10.1007/BF00577166.
2. J.K. Roberts, W.J.C. Orr (1938). "Induced dipoles and the heat of adsorption of argon on ionic crystals". Transactions of the

Faraday Society. 34: 1346. doi:10.1039/TF9383401346.
3. A.M. Sapse, M.T. Rayez-Meaume, J.C. Rayez, L.J. Massa (1979). "Ion-induced dipole H-n clusters". Nature. 278 (5702): 332. doi:10.1038/278332a0.

106 H.-J. Schneider (2015). "Dispersive Interactions in Solution Complexes Dispersive Interactions in Solution Complexes", Acc. Chem. Res 48, 1815-1822.

107 H.C. Hamaker (1937). "The London-van der Waals attraction between spherical particles", Physica, 4(10), 1058-1072 (1937). 10.1016/S0031-8914(37)80203-7.

108 1. E.M. Lifshitz, Soviet Phys. JETP, 2, 73 (1956).
 2. D. Langbein, Phys. Rev. B, 2, 3371 (1970).

109 D.J. Griffiths (1998). Introduction to Electrodynamics (3rd ed.). chapter 12 Prentice Hall. ISBN 978-0-13-805326-0. OCLC 40251748.

110 D.J. Griffiths (1999). Introduction to Electrodynamics (3rd ed.). Prentice Hall. p. 438. ISBN 978-0-13-805326-0. OCLC 40251748. Griffiths 1999, pp. 266-268.

111 R.S. Rao (2012). Microwave Engineering. PHI Learning.

112 Y. Chiang, et al. Physical Ceramics, John Wiley & Sons 1997, New York.

113 Y. Chiang, et al. ibid.

114 Kuhn, U.; Luty, F. (1965). "Paraelectric heating and cooling with OH-dipoles in alkali halides". Solid State Communications. 3 (2): 31. doi:10.1016/0038-1098(65)90060-8.

115 1. W. Kanzig (1957). "Ferroelectrics and Antiferroelectrics". In F. Seitz; T.P. Das; D. Turnbull; E.L. Hahn. Solid State Physics. 4. Academic Press. p. 5. ISBN 978-0-12-607704-9.
 2. M. Lines, A. Glass (1979). Principles and applications of ferroelectrics and related materials. Clarendon Press, Oxford. ISBN 978-0-19-851286-8.

116 J. Valasek (1920). "Piezoelectric and allied phenomena in Rochelle salt". Physical Review. 15 (6): 537. doi:10.1103/PhysRev.15.505.

117 A. Safari (2008). Piezoelectric and acoustic materials for transducer applications. Springer Science & Business Media. p. 21. ISBN 978-0387765402.

118 M. Suzuki, N. Kawamura, H. Miyagawa, J.S. Garitaonandia, Y. Yamamoto, H. Hori (24 Jan 2012). "Measurement of a Pauli and Orbital Paramagnetic State in Bulk Gold Using X-Ray Magnetic Circular Dichroism Spectroscopy". Physical Review Letters. doi:10.1103/PhysRevLett.108.047201. Retrieved 3 Oct 2018.

119 G.H. Lander, D.J. Lam (1976). "Neutron diffraction study of PuP: The electronic ground state". Phys. Rev. B. 14 (9): 4064-67. doi:10.1103/PhysRevB.14.4064.

120 A.T. Aldred, B.D. Dunlap, D.J. Lam, G.H. Lander, M.H. Mueller, I. Nowik (1975). "Magnetic properties of neptunium Laves phases: $NpMn_2$, $NpFe_2$, $NpCo_2$, and $NpNi_2$". Phys. Rev. B. 11 (1): 530-44. doi:10.1103/PhysRevB.11.530.

121 M.H. Mueller, G.H. Lander, H.A. Hoff, H.W. Knott, J.F. Reddy (1979). "Lattice distortions measured in actinide ferromagnets PuP, $NpFe_2$, and $NpNi_2$". J Phys Colloque C4, supplement. 40 (4): C4-68-C4-69.

122 G.-B. Jo, Y.-R. Lee, J.-H. Choi, C.A. Christensen, T.H. Kim, J.H. Thywissen, D.E. Pritchard, W. Ketterle (2009). "Itinerant Ferromagnetism in a Fermi Gas of Ultracold Atoms". Science. 325 (5947): 1521-24. doi:10.1126/science.1177112.

123 S. Chikazumi (2009). Physics of ferromagnetism. English edition prepared with the assistance of C.D. Graham, Jr (2nd ed.). Oxford: Oxford University Press. pp. 129-30. ISBN 9780199564811.

124 Ferromagnetism". University of California, San Diego. Retrieved 2 January 2013

125 F. Yen, R.P. Chaudhury, E Galstyan, B. Lorenz, Y.Q. Wang, Y.Y. Sun, C.W. Chu (2008). "Magnetic phase diagrams of the Kagome staircase compound $Co_3V_2O_8$". Physica B: Condensed Matter. 403 (5-9): 1487-1489. doi:10.1016/j.physb.2007.10.334.

126 M. Forrester, F. Kusmartsev (2014). "The nano-mechanics

and magnetic properties of high moment synthetic antiferromagnetic particles". Physica Status Solidi A. 211 (4): 884-889. doi:10.1002/pssa.201330122.

127 N.A. Spaldin (2010). "9. Ferrimagnetism". Magnetic materials : fundamentals and applications (2nd ed.). Cambridge: Cambridge University Press. pp. 113-129. ISBN 9780521886697.

128 L. Néel (1948). "Propriétées magnétiques des ferrites; Férrimagnétisme et antiferromagnétisme", Annales de Physique (Paris) 3, 137-198.

129 C. Klein, B. Dutrow, Mineral Science, 23rd ed., Wiley, p. 243.

130 C.D. Stanciu, A.V. Kimel, F. Hansteen, A. Tsukamoto, A. Itoh, A. Kirilyuk, and Th. Rasing (2006). Ultrafast spin dynamics across compensation points in ferrimagnetic GdFeCo: The role of angular momentum compensation, Phys. Rev. B 73, 220402(R) (2006).

131 X.-G. Wen, E. Witten (1985). Electric and magnetic charges in superstring models, Nuclear Physics B 261, 651-677.

132 K.A. Milton (2006). "Theoretical and experimental status of magnetic monopoles". Reports on Progress in Physics. 69 (6): 1637-1711. doi:10.1088/0034-4885/69/6/R02.

133 C, Castelnovo, R. Moessner, S.L. Sondhi (2008). "Magnetic monopoles in spin ice". Nature. 451 (7174): 42-45. doi:10.1038/nature06433.

134 M.W. Ray, E. Ruokokoski, S. Kandel, M. Möttönen, D.S. Hall (2014). "Observation of Dirac monopoles in a synthetic magnetic field". Nature. 505 (7485): 657-660. doi:10.1038/nature12954.

135 P. Dirac (1931). "Quantised Singularities in the Electromagnetic Field" Paul Dirac, Proceedings of the Royal Society, May 29, 1931. Retrieved February 1, 2014.

136 1. Y.B. Zel'dovich, M.Y. Khlopov (1978). "On the concentration of relic monopoles in the universe". Phys. Lett. B79 (3): 239-41. doi:10.1016/0370-2693(78)90232-0.
  2. J. Preskill (1979). "Cosmological production of superheavy magnetic monopoles". Phys. Rev. Lett. 43 (19): 1365-1368.

doi:10.1103/PhysRevLett.43.1365.

137 Guth, Alan (1997). The Inflationary Universe: The Quest for a New Theory of Cosmic Origins. Perseus. ISBN 978-0-201-32840-0.

138 1. The magnetism of matter, Feynman Lectures in Physics Ch 34.
    2. Ferromagnetism, Feynman Lectures in Physics Ch 36.

139 C. Kittel (2004). "Introduction to Solid State Physics", (8 ed.) John Wiley & Sons. ISBN 978-0471415268.

140 L.J. Curtis (2003). Atomic Structure and Lifetimes: A Conceptual Approach. Cambridge University Press. p. 74.

141 G.P. Thomson (1927). "Diffraction of Cathode Rays by a Thin Film". Nature. 119 (3007): 890-890.

142 S. Gorfman, O. Schmidt, V. Tsirelson, M. Ziolkowski, U. Pietsch, "Crystallography under external electric field", ZAAC 639 (2013) 1953-1962. https://doi.org/10.1002/zaac.201200497.

143 C Kittel. ibid 138

144 D.V. Schroeder (2000). "An Introduction to Thermal Physics" Addison-Wesley, San Francisco. Section 7.5.

145 R.A. Matula (1979). "Electrical resistivity of copper, gold, palladium, and silver". Journal of Physical and Chemical Reference Data 8, 1147-1298 DOI:10.1063/1.555614.

146 L.E. Kinsler, et al. (2000). Fundamentals of acoustics, 4th Ed., John Wiley and sons Inc., New York, USA.

147 D.T. Queheillalt, H.N.G. Wadley (1998). "Temperature dependence of the elastic constants of solid and liquid $Cd_{0.96}Zn_{0.04}Te$ obtained by laser ultrasound". J. Appl. Phys. 83, 4124-4133.

148 K. Wang, R.R. Reeber (1996). "Thermal Expansion of Copper", High Temperature and Materials Science 35, 181-186.

149 M. Gu, C. Q Sun, Y. Zhou, "An approach of local band average for the temperature dependence of lattice thermal

expansion", www.researchgate.net-publication-1905865.

150 W. Paszkowicz, R. Minikayev, P. Piszora, M. Knapp, C. Bahtz, "Low Temperature Measurements of Lattice Parameter of Microcrystalline Gold, http://hasyweb.desy.de/science/annual_reports/2005_report/part1/contrib/42/15263.pdf.

151 K. Wang, R.R. Reeber, "Thermal Expansion of Copper", High Temperature and Materials Science 35 (1996) 181-186.

152 S. Pandini, A. Pegoretti (2011). "Time and temperature effects on Poisson's ratio of poly(butylene terephthalate)", eXPRESS Polymer Letters 5(8), 685-697, DOI: 10.3144/expresspolymlett.2011.67.

153 www.webelements.com/copper/crystal_structure.html

154 1. E. Merzbacher (1998). Quantum Mechanics (3rd ed.) 372-3
2. D. Griffiths (2005). Introduction to Quantum Mechanics (2nd ed.) 183-4.

155 J.S. Townsend, A modern approach to quantum mechanics, University Science Books. p. 31 and p. 80. ISBN 978-1891389788

156 R. Eisberg, R. Resnick (1985). Quantum Physics of Atoms, Molecules, Solids, Nuclei, and Particles (2nd ed.). pp. 272-3.

157 T.E. Phipps, J.B. Taylor (1927). "The Magnetic Moment of the Hydrogen Atom". Physical Review. 29 (2): 309-320. doi:10.1103/PhysRev.29.309.

158 W. Gerlach, O. Stern (1922). "Der experimentelle Nachweis der Richtungsquantelung im Magnetfeld". Zeitschrift für Physik. 9: 349-352.

159 D. Griffiths (2005). Introduction to Quantum Mechanics (2nd ed.). pp. 183-4.

160 A. Pais (1991). Niels Bohr's Times. Oxford: Clarendon Press. p. 201. ISBN 0-19-852049-2.

161 P.J. Nahin (1992). "Maxwell's grand unification". Spectrum, IEEE. 29 (3): 45.

162 G. Hall (2008). "Maxwell's electromagnetic theory and special

relativity", Phil. Trans. R. Soc. A 366, 1849-1860.

163 Nature News. Archived 27.08.2015.

164 S. Weinberg (2002). Foundations, The Quantum Theory of Fields, I, Cambridge University Press, ISBN 0-521-55001-7.

165 R. Kodama (2016). "Nonlinear interaction of laser light with vacuum", Apollon First User's Meeting 12-13 Feb. 2016.

166 1. R.A. Serway (1990). Physics for Scientists & Engineers (3rd ed.). Saunders. p. 1150. ISBN 0-03-030258-7.
2. F.W. Sears, M.W. Zemansky, H.D. Young (1983). University Physics (6th ed.). Addison-Wesley. pp. 843-844. ISBN 0-201-07195-9.

167 1. P. Lenard (1902). "Über die lichtelektrische Wirkung". Annalen der Physik. 313 (5): 149-198. doi:10.1002/andp.19023130510.
2. R. Millikan (1914). "A Direct Determination of "h."". Physical Review. 4 (1): 73-75. doi:10.1103/PhysRev.4.73.2.
3. R. Millikan (1916). "A Direct Photoelectric Determination of Planck's "h"". Physical Review. 7 (3): 355-388. doi:10.1103/PhysRev.7.355.

168 F. Bubb (1924). "Direction of Ejection of Photo-Electrons by Polarized X-rays". Physical Review. 23 (2): 137-143. doi:10.1103/PhysRev.23.137.

169 R. Gautreau, W. Savin (1999). Schaum's Outline of Modern Physics (2nd ed.). McGraw-Hill. pp. 60-61. ISBN 0-07-024830-3.

170 M. Vranic, O. Klimo, G. Korn, S. Weber (2018). "Multi-GeV electron-positron beam generation from laser-electron scattering", Scientific Reports 8, Article number: 4702.

171 J.Q. Yu, H.Y. Lu, T. Takahashi, R.H. Hu, Z. Gong, W.J. Ma, Y.S. Huang, C.E. Chen, X.Q. Yan (2019). "Creation of Electron-Positron Pairs in Photon-Photon Collisions Driven by 10-PW Laser Pulses" Phys. Rev. Lett. 122, 014802.

172 M. Planck (1914). The Theory of Heat Radiation. Masius, M. (transl.) (2nd ed.). P. Blakiston's Son & Co. OL 7154661M. p42.

173 G. Shao, et al. (2019). "Improved oxidation resistance of high

emissivity coatings on fibrous ceramic for reusable space systems". Corrosion Science. 146: 233-246. doi:10.1016/j.corsci.2018.11.006.

174 B. Stewart (1858). "An account of some experiments on radiant heat". Transactions of the Royal Society of Edinburgh. 22: 1-20. doi:10.1017/S0080456800031288.

175 B. Hapke (1993). Theory of Reflectance and Emittance Spectroscopy. Cambridge University Press, Cambridge UK. ISBN 978-0-521-30789-5. pp. 362-373.

176 R. Loudon (2000). The Quantum Theory of Light (3rd ed.). Oxford University Press. ISBN 978-0-19-850177-0. pp. 3-45.

177 P.F. Dahl (1992). Superconductivity, Its Historical Roots and Development from Mercury to the Ceramic Oxides, American Institute of Physics, New York, p. 13.

178 P.J. Ray. Figure 2.4 in Master's thesis, "Structural investigation of $La_{2-x}Sr_xCuO_{4+y}$ - Following staging as a function of temperature". University of Copenhagen. 2015. DOI:10.6084/m9.figshare.2075680.v2.

179 1. J. Bardeen, L.N. Cooper, J.R. Schrieffer (1957). "Microscopic Theory of Superconductivity". Physical Review. 106 (1): 162-164.
2. J. Bardeen, L.N. Cooper, J.R. Schrieffer (1957). "Theory of Superconductivity". Physical Review. 108 (5): 1175-1205.

180 S. Qin, J. Kim, Q. Niu, C.-K. Shih, Superconductivity at the Two-Dimensional Limit, Science 324 (2009), 1314-1317.

181 V.Z. Kresin and W.A. Little (eds.). Organic Superconductivity. Springer Science+Business Media New York 1990.

182 J.H. Schön, Ch. Kloc, B. Batlogg (2000). Superconductivity at 52 K in hole-doped C60. Nature 408, 549-552.

183 D. Pines (1997). Understanding High Temperature Superconductors: Progress and Prospects, Physica C 282-287, 273 and the references herein.

184 G. Xiao, F.H. Streitz, A. Gavrin, Y.W. Du, C.L. Chien (1987a). Phys. Rev. B 35, 8782.

185 L.S. Chandra, M.K. Chattopadhyay, S.B. Roy, V.C. Sahni, G.R. Myneni (2012). "Magneto thermal conductivity of superconducting Nb with intermediate level of impurity", Supercond. Sci. Technol. 25, 035010.

186 M. Debessai, T. Matsuoka, J.J. Hamlin, W. Bi, Y. Meng, K. Shimizu, J.S. Schilling (2010). J. Phys.: Conf. Series 215, 012034.

187 J.Z. Hu, L.D. Merkel, C.S. Menoni, I.L. Spain (1986). Crystal data for high-pressure phases of silicon, Physical review. B, Condensed matter, 34, 4679-4684, DOI: 10.1103/PhysRevB.34.4679.

188 G. Martinez, J.M. Mignot, G. Chouteau, K.J. Chang, M.M. Dacorogna, M.L. Cohen (1986). Superconductivity of Silicon, Phys. Scr. T13, 226-229.

189 S.V. Vonsovsky, Y.A. Izyumov, E.Z. Kurmaev (1982). Superconductivity of Transition Metals: Their Alloys and Compounds, Springer Series in Solid-State Sciences Vol. 27, Springer-Verlag, Berlin, Heidelberg, New York, p 27.

190 ibid 188, but p 245.

191 C.P. Poole, Jr, H.A. Farach, R.J. Creswick (1995). Superconductivity, Academic Press Inc. Chapter 9. Type II Superconductivity.

192 F.S. Wells, A.V. Pan, X.R. Wang, S.A. Fedoseev, H. Hilgenkamp (2015). "Analysis of low-field isotropic vortex glass containing vortex groups in $YBa_2Cu_3O_{7-x}$ thin films visualized by scanning SQUID microscopy". Scientific Reports. 5: 8677. doi:10.1038/srep08677.

193 M. Tinkham (1996). Introduction to Superconductivity, Second Edition. New York, NY: McGraw-Hill. ISBN 0486435032.

194 A.A. Abrikosov (2003). "Type II superconductors and the vortex lattice", Nobel Lecture, December 8, 2003.

195 1. P.J. Saunders, G.A. Ford (2005). The rise of the superconductors. Boca Raton, Fla.: CRC Press. ISBN 9780748407729.
2. J.G. Bednorz, K.A. Müller (1986). "Possible high TC

superconductivity in the Ba-La-Cu-O system". Zeitschrift für Physik B. 64 (2): 189-193. doi:10.1007/BF01303701.

196 1. A. Leggett (2006). "What DO we know about high T_c?". Nature Physics. 2 (3): 134-136. doi:10.1038/nphys254.
2. Z.-A. Ren, et al. (2008). "Superconductivity and phase diagram in iron-based arsenic-oxides ReFeAsO1-δ (Re=rare-earth metal) without fluorine doping". EPL. 83 (1): 17002. doi:10.1209/0295-5075/83/17002.

197 1. A.P. Drozdov, M.I. Eremets, I.A. Troyan, V. Ksenofontov, S.I. Shylin (2015). "Conventional superconductivity at 203 kelvin at high pressures in the sulfur hydride system". Nature. 525 (7567): 73-6. doi:10.1038/nature14964.
2. H, Christoph, B. Lilia (2015). "Influence of bonding on superconductivity in high-pressure hydrides". Phys Rev B. 92 (6): 060508. doi:10.1103/PhysRevB.92.060508.

198 R. Hazen, L. Finger, R. Angel, C. Prewitt, N. Ross, H. Mao, C. Hadidiacos, P. Hor, R. Meng, C. Chu (1987). "Crystallographic description of phases in the Y-Ba-Cu-O superconductor". Physical Review B. 35 (13): 7238-7241. doi:10.1103/PhysRevB.35.7238.

199 N.N. Greenwood, A. Earnshaw (1997). Chemistry of the Elements (2nd ed.). Butterworth-Heinemann. ISBN 0-08-037941-9.

200 N. Khare (2003). Handbook of High-Temperature Superconductor Electronics. CRC Press. ISBN 978-0-8247-0823-8.

201 C. Hartinger. "DFG FG 538 - Doping Dependence of Phase transitions and Ordering Phenomena in Cuprate Superconductors". Wmi.badw-muenchen.de. Archived from the original on 27 December 2008.

202 A.M. Hermann, J.V. Yakhmi eds. (1994). Thallium-Based High-Temperature Superconductors, Marcel Dekker.

203 R. Hazen, et al. (1988). "Superconductivity in the high-T_c Bi-Ca-Sr-Cu-O system: Phase identification". Physical Review Letters. 60 (12): 1174-1177. doi:10.1103/PhysRevLett.60.1174.

204 J. Tarascon, W. McKinnon, P. Barboux, D. Hwang, B. Bagley, L. Greene, G. Hull, Y. Lepage, N. Stoffel, M. Giroud (1988). "Preparation, structure, and properties of the superconducting compound series $Bi_2Sr_2Ca_{n-1}Cu_nO_y$ with n=1, 2, and 3". Physical Review B. 38 (13): 8885-8892. doi:10.1103/PhysRevB.38.8885.

205 Z.Z. Sheng, A.M. Hermann, A. El Ali, C. Almasan, J. Estrada, T. Datta, R.J. Matson (1988). "Superconductivity at 90 K in the Tl-Ba-Cu-O system". Physical Review Letters. 60 (10): 937-940. doi:10.1103/PhysRevLett.60.937.

206 Z.Z. Sheng, A.M. Hermann, (1988). "Superconductivity in the rare-earth-free Tl-Ba-Cu-O system above liquid-nitrogen temperature". Nature. 332 (6159): 55-58. doi:10.1038/332055a0.

207 A. Schilling, M. Cantoni, J.D. Guo, H.R. Ott, (1993). "Superconductivity above 130 K in the Hg-Ba-Ca-Cu-O system". Nature. 363 (6424): 56-58. doi:10.1038/363056a0.

208 C.W. Chu, L. Gao, F. Chen, Z.J. Huang, R.L. Meng, Y.Y. Xue (1993). "Superconductivity above 150 K in $HgBa_2Ca_2Cu_3O_{8+\delta}$ at high pressures". Nature. 365 (6444): 323-325. doi:10.1038/365323a0.

209 J.-F. Ge, et al. (2014). "Superconductivity in single-layer films of FeSe with a transition temperature above 100 K". Nature Materials. 1406 (3): 285-9. doi:10.1038/nmat4153.

210 Z.-A. Ren, et al. (2008). "Superconductivity and phase diagram in iron-based arsenic-oxides $ReFeAsO_{1-\delta}$ (Re=rare-earth metal) without fluorine doping". EPL. 83 (1): 17002. arXiv:0804.2582.

211 1. M. Rotter, M. Tegel, D. Johrendt (2008). "Superconductivity at 38 K in the Iron Arsenide (Ba1-xKx)Fe2As2". Physical Review Letters. 101 (10): 107006. doi:10.1103/PhysRevLett.101.107006.
  2. K. Sasmal, B. Lv, B. Lorenz, A.M. Guloy, F. Chen, Y.Y. Xue, C.W. Chu, (2008). "Superconducting Fe-Based Compounds (A1-xSrx)Fe2As2 with A=K and Cs with Transition Temperatures up to 37 K". Physical Review Letters. 101 (10): 107007. doi:10.1103/PhysRevLett.101.107007.

212 1. M.J. Pitcher, D.R. Parker, P. Adamson, S.J. Herkelrath, A.T. Boothroyd, R.M. Ibberson, M Brunelli, S.J. Clarke (2008). "Structure and superconductivity of LiFeAs". Chemical

Communications. 2008 (45): 5918-5920. doi:10.1039/b813153h.
2. J.H. Tapp, Z. Tang, B. Lv, K. Sasmal, B. Lorenz, P.C.W. Chu, A.M. Guloy (2008). "LiFeAs: An intrinsic FeAs-based superconductor with T_c=18 K". Physical Review B. 78 (6): 060505. doi:10.1103/PhysRevB.78.060505.
3. D.R. Parker, M.J. Pitcher, P.J. Baker, I. Franke, T. Lancaster, S.J. Blundell, S.J. Clarke (2009). "Structure, antiferromagnetism and superconductivity of the layered iron arsenide NaFeAs". Chemical Communications. 2009 (16): 2189-2191. doi:10.1039/b818911 K.

213 J. Zhao, Q. Huang, C. de la Cruz, S. Li, J.W. Lynn, Y. Chen, M.A. Green, G.F. Chen, G. Li, Z. Li, J.L. Luo, N.L. Wang, P. Dai (2008). "Structural and magnetic phase diagram of CeFeAsO1−xFx and its relation to high-temperature superconductivity". Nature Materials. 7 (12): 953-959. doi:10.1038/nmat2315.

214 A.A. Kordyuk (2012). "Iron-based high temperature superconductors: Magnetism, superconductivity, and electronic structure (Review Article)". Low Temp. Phys. 38 (9): 888-899. doi:10.1063/1.4752092.

215 H. Luetkens, et al. (2009). "Electronic phase diagram of the LaO1-xFxFeAs superconductor". Nature Materials. 8 (4): 305-9. doi:10.1038/nmat2397.

216 1. A.J Drew, et al. (2009). "Coexistence of static magnetism and superconductivity in SmFeAsO1-xFx as revealed by muon spin rotation". Nature Materials. 8 (4): 310-314. doi:10.1038/nmat2396.
2. S. Sanna, R. De Renzi, G. Lamura, C. Ferdeghini, A. Palenzona, M. Putti, M. Tropeano, T. Shiroka (2009). "Competition between magnetism and superconductivity at the phase boundary of doped SmFeAsO pnictides". Physical Review B. 80 (5): 052503. doi:10.1103/PhysRevB.80.052503.

217 J. Zhao, et al. (2008). "Structural and magnetic phase diagram of $CeFeAsO_{1-x}F_x$ and its relation to high-temperature superconductivity". Nature Materials. 7 (12): 953-959. doi:10.1038/nmat2315.

218 J.-H. Chu, J. Analytis, C. Kucharczyk, I. Fisher (2009). "Determination of the phase diagram of the electron doped

superconductor Ba(Fe$_{1-x}$Co$_x$)$_2$As$_2$". Physical Review B. 79 (1): 014506. doi:10.1103/PhysRevB.79.014506.

219 C.-H. Lee, A. Iyo, H. Eisaki, H. Kito, M. Teresa Fernandez-Diaz, T. Ito, K. Kihou, H. Matsuhata, M. Braden, K. Yamada (2008). "Effect of Structural Parameters on Superconductivity in Fluorine-Free LnFeAsO$_{1-y}$ (Ln=La, Nd)". Journal of the Physical Society of Japan. 77 (8): 083704. doi:10.1143/JPSJ.77.083704.

220 A. Drozdov, M.I. Eremets, I.A. Troyan (2014). "Conventional superconductivity at 190 K at high pressures". arXiv:1412.0460.

221 Y.F. Ge, F. Zhang, Y.G. Yao (2016). "First-principles demonstration of superconductivity at 280 K in hydrogen sulfide with low phosphorus substitution". Phys. Rev. B. 93 (22): 224513. doi:10.1103/PhysRevB.93.224513.

222 A. Leggett (2006). "What DO we know about high Tc?". Nature Physics. 2 (3): 134-136. doi:10.1038/nphys254.

223 T. Timusk, S. Bryan (1999). "The pseudogap in high-temperature superconductors: an experimental survey". Reports on Progress in Physics. 62 (1): 61-122. doi:10.1088/0034-4885/62/1/002.

224 N. Mannella, et al. (2005). "Nodal quasiparticle in pseudogapped colossal magnetoresistive manganites". Nature. 438 (7067): 474-478. doi:10.1038/nature04273.

225 P. Monthoux, A. Balatsky, D. Pines (1992). "Weak-coupling theory of high-temperature superconductivity in the antiferromagnetically correlated copper oxides". Physical Review B. 46 (22): 14803-14817. doi:10.1103/PhysRevB.46.14803.

226 S. Chakravarty, A. Sudbø, P.W. Anderson, S. Strong (1993). "Interlayer Tunneling and Gap Anisotropy in High-Temperature Superconductors". Science. 261 (5119): 337-340. doi:10.1126/science.261.5119.337.

227 J. Phillips (2010). "Percolative theories of strongly disordered ceramic high-temperature superconductors". Proceedings of the National Academy of Sciences of the United States of America. 43 (4): 1307-10. doi:10.1073/pnas.0913002107.

228 V. Geshkenbein, A. Larkin, A Barone (1987). "Vortices with half magnetic flux quanta in heavy-fermion superconductors". Physical Review B. 36 (1): 235-238. doi:10.1103/PhysRevB.36.235.

229 J.R. Kirtley, C.C. Tsuei, J.Z. Sun, C.C. Chi, L.S. Yu-Jahnes, A. Gupta, M. Rupp, M.B. Ketchen (1995). "Symmetry of the order parameter in the high-Tc superconductor $YBa_2Cu_3O_{7-\delta}$". Nature. 373 (6511): 225-228. doi:10.1038/373225a0.

230 J.R. Kirtley, C.C. Tsuei, A. Ariando, C.J.M. Verwijs, S. Harkema, H. Hilgenkamp (2006). "Angle-resolved phase-sensitive determination of the in-plane gap symmetry in $YBa_2Cu_3O_{7-\delta}$". Nature Physics. 2 (3): 190-194. doi:10.1038/nphys215.

231 C.C. Tsuei, J.R. Kirtley, Z.F. Ren, J.H. Wang, H. Raffy, Z.Z. Li (1997). "Pure $d_{x^2-y^2}$ order-parameter symmetry in the tetragonal superconductor $Tl_2Ba_2CuO_{6+\delta}$". Nature. 387 (6632): 481-483. doi:10.1038/387481a0.

232 S.J. Hagen, T.W. Jing, Z.Z. Wang, J. Horvath, N.P. Ong (1988). Phys. Rev. B37, 7928.

233 T. Sekitani, N. Miura, S. Ikeda, Y.H. Matsuda, Y. Shiohara (2004). "Upper critical field for optimally-doped $YBa_2Cu_3O_{7-\delta}$". Physica B: Condensed Matter. 346-347: 319.

234 M. Kriener, et al. "Evolution of electronic states and emergence of superconductivity in the polar semiconductor GeTe by doping valence-skipping In", ResearchGate publication 330673125.

235 S.K. Shrivastava (2018). "Crystal Structure of Cuprate based Superconducting Materials", Inter. J. Eng. Sci. Math. 7, 151-159.

236 B. Wang, K. Ohgushi, "Superconductivity in anti-post-perovskite vanadium compounds", SCIENTIFIC REPORTS | 3 : 3381 | DOI: 10.1038/srep03381.

- 299 -

www.ingramcontent.com/pod-product-compliance
Lightning Source LLC
Chambersburg PA
CBHW052342220526
45465CB00003BA/924